META-ANALYSIS IN PSYCHIATRY RESEARCH

Fundamental and Advanced Methods

META-ANALYSIS IN PSYCHIATRY RESEARCH

Fundamental and Advanced Methods

Mallikarjun B. Hanji, PhD

Apple Academic Press Inc.	Apple Academic Press Inc.
3333 Mistwell Crescent	9 Spinnaker Way
Oakville, ON L6L 0A2	Waretown, NJ 08758
Canada	USA

©2017 by Apple Academic Press, Inc.

First issued in paperback 2021

Exclusive worldwide distribution by CRC Press, a member of Taylor & Francis Group
No claim to original U.S. Government works

ISBN 13: 978-1-77463-623-7 (pbk)
ISBN 13: 978-1-77188-376-4 (hbk)

Library and Archives Canada Cataloguing in Publication

Hanji, Mallikarjun B., author
Meta-analysis in psychiatry research : fundamental
and advanced methods / Mallikarjun B. Hanji, PhD.

Includes bibliographical references and index.
Issued in print and electronic formats.
ISBN 978-1-77188-376-4 (hardcover).--ISBN 978-1-315-36623-4 (PDF)
1. Meta-analysis. 2. Psychiatry--Research. I. Title.

| R853.M48H35 2017 | 616.89'00727 | C2016-907985-6 | C2016-907986-4 |

Library of Congress Cataloging-in-Publication Data

Names: Hanji, Mallikarjun B., author.
Title: Meta-analysis in psychiatry research : fundamental and advanced methods / author, Mallikarjun B. Hanji.
Description: Toronto ; New Jersey : Apple Academic Press, 2017. | Includes bibliographical references and index.
Identifiers: LCCN 2016055226 (print) | LCCN 2016056951 (ebook) | ISBN 9781771883764 (hardcover : alk. paper) | ISBN 9781315366234 (ebook)
Subjects: LCSH: Meta-analysis. | Psychiatry--Methodology. | Psychiatry--Reseach. | MESH: Psychometrics--methods | Research Design | Meta-Analysis as Topic
Classification: LCC R853.M48 M467 2017 (print) | LCC R853.M48 (ebook) | NLM BF 39 | DDC 616.890072--dc23
LC record available at https://lccn.loc.gov/2016055226

Apple Academic Press also publishes its books in a variety of electronic formats. Some content that appears in print may not be available in electronic format. For information about Apple Academic Press products, visit our website at **www.appleacademicpress.com** and the CRC Press website at **www.crc-press.com**

ABOUT THE AUTHOR

Mallikarjun B. Hanji, PhD

Mallikarjun B. Hanji is presently working as a Chief Technical Officer at the Agricultural Technology Application Research Institute, Bangalore, India, and has over 25 years of research, teaching, and extension experience. He has published more than five scientific papers, including an introductory paper on meta-analysis for mental health care research followed by a pattern of prevalence of mental retardation and affective disorders in India. His main contribution is the identification of suitable methods for estimation of pattern and prevalence for schizophrenia and epilepsy in India. He received his MSc in agricultural statistics from the University of Agricultural Sciences Bangalore; an MBA from Indira Gandhi National Open University, New Delhi; and a PhD in Biostatistics (topic: Meta-analytical Approach to Estimate Pattern of Prevalence of Schizophrenia and Epilepsy in India) from the Department of Biostatistics, National Institute of Mental Health and Neurosciences (NIMHANS), Bangalore, India.

CONTENTS

LIST OF ABBREVIATIONS

CI	confidence interval
DL method	DerSimonian and Laird method
DOR	diagnostic odds ratio
DSM	Diagnostic and Statistical Manual
FN	false negatives
FP	false positives
ICD	International Classification of Diseases
ICMR	Indian Council of Medical Research
IPD	individual patient data meta-analysis
IPSS	Indian Psychiatric Survey Schedule
ML	maximum likelihood
MLE	maximum likelihood estimation
NIMHANS	National Institute of Mental Health and Neurosciences
NNT	number needed to treat
OR	odds ratio
QUOROM	quality of reporting of meta-analyses
RD	risk difference
REML	restricted maximum likelihood estimate
RevMan	review manager
RPES	Rapid Psychiatric Evaluation Scheme
RR	risk ratio
SE	standard error
TN	true negatives
TR	true positives
WHO	World Health Organization

PREFACE

This book is best suited for professionals, teachers, and post-graduate students in the field of psychiatry and allied fields. The main objective of writing this book is to introduce the latest meta-analytical methods developed and applications of suitable ones in the field of psychiatry with real examples in estimate pattern and prevalence of schizophrenia in India along with review of software to be used for the same in a precise and simple manner. The book contains most of the methods developed in meta-analysis, which are described in simple language and presented in a systematic and chronological order so that reader can easily understand the importance of individual methods.

Review of software: The software to be used for meta-analysis has been reviewed in a systematic way to assist the reader in choosing the required software. The commands of the software, namely, STATA, have been used extensively to demonstrate the examples in detail.

ACKNOWLEDGMENTS

I am grateful to Dr. M. Venkaswamy Reddy, retired senior faculty, Department of Biostatistics, National Institute of Mental Health and Neurosciences (NIMHANS), Bangalore, India, for scrutinizing the manuscript and providing valuable suggestions. I thank Ashish Kumar of Apple Academic Press for his full cooperation in bringing this book in its present form.

—**Mallikarjun B. Hanji**

CHAPTER 1

INTRODUCTION

CONTENTS

ABSTRACT

Meta-analysis can be defined as a systematic statistical method for analyzing and synthesizing results from independent studies, taking into account all pertinent information. Readers of narrative studies face problems such as lack of detailed description, the process that led to the review, and hence the readers cannot replicate and verify the results and conclusions of the review. Most effective mechanism for systematic review is to reduce bias and increase precision, by including maximum possible number of relevant individual studies and providing a detailed description of their strengths and limitations. Vote counting is clearly unsound, since it ignores sample size, effect size, and research design. Meta-analysis is trying to answer four basic questions, namely, (1) are the results of the different studies similar and to the extent that they are similar, (2) what is the best overall estimate, (3) how precise and robust is the estimate, and (4) can dissimilarities be explained. Exploratory analysis, such as regarding subgroups of patients who are likely to respond particularly well to a treatment, may generate promising new research questions. Meta-analysis identifies areas where further studies are needed. Meta-analysis provides robust evidence and may utilize a less biased sample of evidence. Physicians can now make decisions regarding the use of therapies or diagnostic procedures on the basis of a single article that synthesizes the findings of tens or hundreds of clinical studies. The Cochrane Collaboration which is an international organization involved in preparing meta-analysis of the effects of interventions in all aspects of health care. The science of meta-analysis is relevant to clinical and community psychiatry to evaluate the potential errors and sources of bias and offer guidelines for evaluation. The statistical basis of meta-analysis reached back to the 17th century wherein astronomy and geodesy intuition and experience suggested that combinations of data might be better than attempts to choose amongst them. Meta-analysis has had critics and criticisms over the years. Most prominent of which is publication bias, which refers to the tendency for journals and authors not to publish articles on research that has no significant findings. There is a danger that meta-analysis of observational data produce very precise but spurious results. The complex methods used in meta-analysis should always be complemented by clinical acumen and common sense in designing the protocol of a systematic review, deciding what data can be combined, and determining whether data should be combined.

Meta-analysis provides an opportunity for shared subjectivity in reviews rather than true objectivity. Meta-analyses are most easily performed with the assistance of computer databases and statistical software.

1.1 FEATURES OF META-ANALYSIS

1.1.1 META-ANALYSIS

Meta-analysis can be defined as a systematic statistical method for analyzing and synthesizing results from independent studies, taking into account all pertinent information. By synthesizing, scrutinizing, tabulating, and perhaps integrating all relevant studies, meta-analysis allows a more objective appraisal, which can help to resolve uncertainties when the original research, classical reviews, and editorial comments disagree. Meta-analysis is a scientific activity that borrows from both the expert review and the methodology of multicenter studies (Fisher et al., 1993). There are varieties of synonyms for meta-analysis used in the literature: overviews, aggregates, syntheses, integration, amalgamation, pooling, and combining. Quantitative is the heart of the meta-analysis and combining results is an essential ingredient in meta-analysis.

1.1.2 NARRATIVE STUDIES

Traditionally, individuals often considered experts in the field who have conducted narrative reviews of the literature, associated with a particular field using informal and subjective methods to collect and interpret information. Readers of narrative studies face problems such as lack of detailed description, the process that led to the review, and hence the readers cannot replicate and verify the results and conclusions of the review.

1.1.3 SYSTEMATIC REVIEWS

Reviews being the product of a scientific process to reduce bias, to increase precision and by providing detailed information to allow replication by others. Most effective mechanism for systematic review is to reduce bias and increase precision, by including maximum possible number of relevant individual studies and providing a detailed description of their strengths and limitations.

1.1.4 VOTE COUNTING METHODS

Once a set of studies have been assembled, a common way to review the results is to count the number of studies reporting various sides of an issue and to choose the view receiving the most votes. This procedure is clearly unsound, since it ignores sample size, effect size, and research design.

1.2 SCOPE AND BENEFITS OF META-ANALYSIS

1.2.1 COMBINE RESULTS

A quantitative systematic review or meta-analysis use statistical methods to combine the results of multiple studies.

1.2.2 HETEROGENEITY

They are trying to answer four basic questions, namely: (1) Are the results of the different studies similar and to the extent that they are similar? (2) What is the best overall estimate? (3) How precise and robust is the estimate? and (4) Can dissimilarities be explained (Lau et al., 1997)?

1.2.3 EXPLORATORY ANALYSIS

Exploratory analysis, such as regarding subgroups of patients who are likely to respond particularly well to a treatment, may generate promising new research questions to be addressed in future studies. Meta-analysis can help us to investigate the relationship between study features and study outcomes. One can code the study features according to the objectives of the review and transform the study outcomes to a common metric so that comparison of the outcome is possible.

1.2.4 IDENTIFICATION OF RESEARCH AREAS

Meta-analysis may demonstrate the level of adequate evidence and this identifies areas where further studies are needed.

1.2.5 PROVIDING EVIDENCE

Meta-analysis can examine questions, provide formal standard of rigorous for accumulating evidence from different studies, formulize the process of policy making, increase statistical power, provide robust evidence, and may utilize a less biased sample of evidence.

Meta-analysis, if appropriate, will enhance the precision of estimates of treatment effects, leading to reduced probability of false negative results, and potentially timely introduction of effective treatments.

1.2.6 BENEFITS OF META-ANALYSIS

Physicians can now make decisions regarding the use of therapies or diagnostic procedures on the basis of a single article that synthesizes the findings of tens or hundreds of clinical studies. Scientists in every field can similarly gain a coherent view of the central reality behind the multifarious and often discordant findings of research in their areas. Meta-analysis of a series of small clinical trials of a new therapy often yields a finding on the basis of which physicians can confidently begin using it without waiting long years for a massive trial to be conducted.

1.3 SOME EXAMPLES

Sharma et al. (2003) has successfully employed meta-analytical procedures to determine the effect of inhaled steroids on bone mineral density. Shann (1997) has employed meta-analysis to obtain evidence of trials of prophylactic antibiotics for children with measles for adequate evidence. The meta-analysis (Gupta and Gupta, 1996; Gupta, 1997) was performed to determine the time trend in the prevalence of coronary heart diseases in India and age and gender specific changes.

The Cochrane Collaboration which is an international organization involved in preparing maintaining and disseminating highly structured, frequently updated, and good quality systematic reviews and meta-analysis of the effects of interventions in all aspects of health care (Cochrane Injuries Group Albumin Reviewer, 1998; Kennedy et al., 2002; Olsen and Gotzsche, 2001).

The national library of medicine defines meta-analysis as a quantitative method of combining the results of independent studies and synthesizing summaries and conclusions, which may be used to evaluate therapeutic effectiveness, plan new studies, etc. with application chiefly in the areas of research and medicine.

Meta-analyses are based on trials of parallel group design, but some trials assessing the treatment of interest may use other designs. This is particularly the case in certain chronic diseases whose treatment is often evaluated by cross over-trials; typical examples include hypertension, asthma, or rheumatic diseases. Parallel and cross-over trials both provide estimates of the same treatment effect (Curtin et al., 2002a,b).

Laird and Ware (1982) have discussed the random effects model for longitudinal data on health effects of air pollution. Malhotra et al. (2001) have conducted a meta-analysis of controlled clinical trials comprising low-molecular-weight heparins with unfractionated heparin in unstable angina. Pavia et al. (2003) have carried out a meta-analysis of residential exposure to radon gas and lung cancer. Ezzat et al. (2004) have carried out a systematic review on the prevalence of pituitary adenomas. Gisbert et al. (2003) have carried out a systematic review and meta-analysis to determine prevalence of hepatitis C virus infection in porphyria cutaneatarda. Devereaux et al. (2002) have carried out meta-analysis of studies comprising mortality rates of private for-profit and private for nonprofit hospitals.

1.3.1 PSYCHIATRIC RESEARCH

The science of meta-analysis is relevant to clinical and community psychiatry to evaluate the potential errors and sources of bias and offer guidelines for evaluation. Meta-analysis is a specific technique that was developed in social sciences, but was soon adapted as a fundamental tool in psychiatric research with a number of aims.

The relevance of meta-analysis to psychiatry stems from one of the earliest meta-analyses ever undertaken, which evaluated efficiency of various forms of psychotherapy. Since the 1980s, meta-analysis has increasingly appeared in the medical literature, and scarcely a month now passes without the publication of a meta-analysis of relevance to clinical psychiatry in general medical journals or in mainstream psychiatric literature (Tharyan, 1998).

Whitehead (1997) has applied a prospectively planned cumulative meta-analysis to a series of concurrent clinical trials. Meta-analysis permits investigation of generalizability and consistency, improved transparency of methodology, and enhance reproducibility in psychiatry fields.

Harrison et al. (2003) have carried out a meta-analysis to answer the question whether brain weight is decreased in schizophrenia patients and concluded that the brain weight is slightly, but significantly, reduced in schizophrenia, consistent in duration and magnitude with MRI volumetric findings.

Based on fitting a model to the funnel plot, Shi and Copas (2004) have discussed a method for random-effects sensitivity analysis that deal with the problems of heterogeneity and publication bias and applied on the effect of alcohol on the risk of breast cancer. Hall and Roter (2002) have conducted a meta-analysis to answer a question: Do patients talk differently to male and female physicians. Reynolds et al. (2003) have carried out a meta-analysis and concluded that heavy alcohol consumption increases the relative risk of stroke while light or moderate alcohol consumption may be protective against total and ischemic stroke.

Ananth et al. (1999) have applied meta-analysis of observational studies on incidence of placental abruption in relation to cigarette smoking and hypertensive disorders during pregnancy and concluded an increased associationship.

Herbert and Cohen (1993) have conducted a meta-analysis and concluded that clinical depression was associated with several large alterations in cellular immunity.

The meta-analytical approaches have wide applications in making diagnosis, deciding on the course and method of treatment, predicting the outcome of treatment, and determining the course of mental disorders in order to prevent them.

1.4   HISTORICAL BACKGROUND

The statistical basis of meta-analysis reached back to the seventeenth century wherein astronomy and geodesy intuition and experience suggested that combinations of data might be better than attempts to choose amongst them.

In 1904: Professor Karl Pearson reported the use of formal techniques to combine data from different studies. G. V. Glass set up a process for synthesizing research studies that used statistical methods, including the use of probabilities and effect sizes, for aggregating results.

Late 1970: Two other coherent methods have been formulated as elaborations of Glass's approach. These five separate and coherent methods are Glassian meta-analysis, study effect meta-analysis, combined probability meta-analysis, meta-analysis using approximate data pooling with tests of homogeneity, and meta-analysis using approximate data pooling with sampling error correction. They indicate the present moment in the continuing evolution of review methodology and can be distinguishable on four factors: purpose, unit of analysis, treatment of study variation, and products of the meta-analysis (Glass, 2000).

In 1976: The same year, that Glass (1976) first coined the term "Meta-analysis," Rosenthal published his book "experimental effects in behavioral research," and Schmidt and Hunter were working on a validity generalization technique. These three concurrent efforts established three distinguishable meta-analytic approaches.

In 1984: Hedges and Olkin (1984) have extended the logic of non-parametric estimators of effect sizes in meta-analysis.

In 1989: Alexander et al. (1989) have developed statistical and empirical examination of the chi-square test for homogeneity of correlations in meta-analysis. From the statistical point of view, meta-analysis is a straight forward application of multifactorial methods (Blend, 2000).

In 1990: The foundation of Cochrane Collaboration facilitated numerous developments (Egger et al., 2001). Researchers have answered the difficulty by supporting methods to test the statistical significance of results combined from separate experiments. They sought ways to combine probability values from tests of significance.

In 1995: Stewart et al. (1995) have conducted a meta-analysis of published studies to identify factors, which explained variation in estimates of migraine prevalence.

In 1997: Todd (1997) has investigated the effect of incorporating one or more sequential trials into a meta-analysis otherwise consisting of fixed sample size trials.

In 2002: Satagopan et al. (2002) have applied meta-analysis in the estimation of cure, relapse, and success rates of short-course chemotherapy in the treatment of pulmonary tuberculosis.

In 2005: Reddy and Hanji (2005) have reviewed the application of meta-analysis in mental health care research.

Meta-analysis is an increasingly popular method for conducting a research review. Because of its quantitative basis, meta-analysis forces reviewers to make explicit a range of decisions that might pass unnoticed in traditional reviews. In exchange, meta-analysis makes possible a more precise characterization of a research domain.

A typical meta-analytic package consists of techniques for (a) combining probabilities across studies, (b) estimating average effect size, (c) determining the stability of results, and (d) identifying factors that moderate the outcomes of separate studies. The application of these techniques requires careful attention to a number of potential problems, including biased selection of studies, inadequacies in the studies comprising the database for the review, and violations of the assumptions of meta-analytic statistical procedures. Notwithstanding these problems, a meta-analysis can advance both theory and application because of its descriptive, diagnostic, predictive, and generative functions (Strube and Hartmann, 1983).

1.4.1 COCHRANE COLLABORATION

The Cochrane Collaboration was founded in 1993 under the leadership of Iain Chalmers. It was developed in response to Archie's call for up-to-date, systematic reviews of all relevant randomized controlled trials of health care. Cochrane's suggestion that the methods used to prepare and maintain reviews of controlled trials in pregnancy and childbirth should be applied more widely was taken up by the Research and Development Programme, initiated to support the United Kingdom's National Health Service.

In October 1995, the Collaboration formed the Cochrane Consumer Network to incorporate patient perspectives into the review process. Shortly thereafter, new "plain language summaries" provided users with a jargon-free synopsis of each systematic review. The Cochrane Collaboration is currently concentrating on capacity building in health research is individuals, groups, and institutions in low- and middle-income countries.

The collaboration formed an official relationship in January 2011 with the World Health Organization (WHO) as a partner nongovernmental organization with a seat on the World Health Assembly to provide input into WHO resolutions. The collaboration is active in providing evidence for good practice during disaster relief and humanitarian crisis through a partnership with Evidence Aid.

The collaboration facilitates in organizing medical research information in a systematic way according to the principles of evidence-based medicine as per the requirement of the health professionals, patients, policy makers, and others in health interventions. The group conducts systematic reviews of randomized controlled trials of health-care interventions, and publishes in The Cochrane Library.

1.5   LIMITATIONS OF META-ANALYSIS

1.5.1   PUBLICATION BIAS

Meta-analysis has had critics and criticisms over the years (Eysenck, 1965). Like most methods, it is not problem-free. Several biases and sources of error have been identified and debated in the literature, most prominent of which is publication bias. Publication bias refers to the tendency for journals and authors not to publish articles on research that has no significant findings. Since the reliability of research synthesis rests on all the effects, including those of no significance, this bias is a vital threat to the method. This bias, however, has received quite a bit of attention (Sterne et al., 2001).

1.5.2   GENERAL PROBLEMS

Walker (1999), after recognizing the current problems with meta-analysis, answered his own question as to whether meta-analysis is really needed. Yes, because there is no serious alternative for taming medical publication. Despite the problems it shares with most methods, meta-analysis has become a well-established and accepted methodology (Moher et al., 1999).

1.5.3   WEAKNESS

The systematic differences in meta-analysis have been largely over looked. It is time they are clarified so the limitations of this approach to research integration can be more realistically assessed. These differences should not be taken as evidence of some inherent weakness of meta-analysis. It is merely a reflection of the natural evolution of a new social scientific tool. It is rooted in the fundamental values of the scientific enterprise: replicability, quantification, causal, and correlational analysis. Valuable information is needlessly scattered in individual studies. The ability of social scientists to deliver generalizable answers to basic questions of policy is, too, serious a concern to allow us to treat research integration lightly. The potential benefits of meta-analysis method seem enormous (Bangert, 1986).

1.5.4   SPURIOUS RESULTS

There is a danger that meta-analysis of observational data produce very precise but spurious results. The statistical combination of data should therefore not be a prominent component of meta-analysis of observational studies. More is gained by carefully examining possible sources of heterogeneity between the results from observational studies. When the purpose of a meta-analysis is to provide estimates of specific effects, then the criteria for inclusion would be more restrictive than if the objectives were to model sources of variability. It is not an easy task and requires careful thought and planning to provide accurate and useful information (Klassen et al., 1998).

1.5.5   NEED COMMON SENSE

The complex methods used in meta-analysis should always be complemented by clinical acumen and common sense in designing the protocol of a systematic review, deciding what data can be combined, and determining whether data should be combined.

1.5.6   NOT EASY

In observational studies that seek information on disease determinants the use of a summary risk ratio may be contentious. The evidence available from observational studies on the casual connection between an exposure and a disease can only be interpreted if much is known about selection, confounding, and measurement bias. That it may not be possible to adequately take into account such biases in any individual study nor in a consistent way across studies (Dwyer et al., 2001; Egger et al., 1998; Stroup et al., 2000). Meta-analyses are neither quick nor easy (Berman and Parker, 2002). Meta-analysis is an important contribution to research and practice but it is not a panacea (Naylor, 1997).

1.5.7   MORE SUBJECTIVE

Meta-analysis provides an opportunity for shared subjectivity in reviews, rather than true objectivity. Authors of meta-analyses must sometimes make decisions based on their own judgment, such as when defining the boundaries of the analysis or deciding exactly how to code moderator variables. However, meta-analysis requires that these decisions are made public so they are open to criticism from other scholars. Meta-analysis is regarded as objective by its proponents but really is subjective. Meta-analysis relies on shared subjectivity rather than objectivity. While every analysis requires certain subjective decisions, these are always stated explicitly so that they are open to criticism.

1.5.8   USE OF COMPUTERS

Meta-analyses are most easily performed with the assistance of computer databases (Microsoft Access, Paradox) and statistical software (DSTAT, SAS).

1.5.9   FAIL TO REPORT

Narrative reviews are not well suited for analyzing the impact of moderating variables. Authors of narrative reviews rarely reach clear conclusions regarding how methodological variations influence the strength of an effect. They also typically fail to report the rules they use to classify studies when looking for the effect of a moderating variable.

KEYWORDS

- narrative studies
- systematic reviews
- vote counting

REFERENCES

Alexander, R. A.; Scozzaro, M. J.; Borodkin, L. J. Statistical and Empirical Examination of the Chi-square Test for Homogeneity of Correlations in Meta-analysis. *Psychol. Bull.* **1989,** *106,* 329–331.

Ananth, C. V.; Smulian, J. C.; Vintzileos, A. M. Incidence of Placental Abruption in Relation to Cigarette Smoking and Hypertensive Disorders during Pregnancy: A Meta-analysis of Observational Studies. *Obstet. Gynecol.* **1999,** *93,* 622–628.

Bangert-Drowns, R. L. Review of Developments in Meta-analytic Method. *Psychol. Bull.* **1986,** *99,* 388–399.

Berman, N. G.; Parker, R. A. Meta-analysis—Neither Quick nor Easy. *BMC Med. Res. Methodol.* **2002,** *2,* 10–18.

Blend, M. *An Introduction to Medical Statistics.* Oxford University Press: New York, 2000.

Cochrane Injuries Group Albumin Reviewers. Human Albumin Administration in Critically Ill Patients—Systematic Review of Randomized Controlled Trials. *Br. Med. J.* **1998,** *317,* 235–240.

Curtin, F.; Altman, D. G.; Elbourne, D. Meta-analysis Combining Parallel and Cross-over Clinical Trials: I Continuous Outcomes. *Stat. Med.* **2002a,** *21,* 2131–2144.

Curtin, F.; Elbourne, D.; Altman, D. G. Meta-analysis Combining Parallel and Cross-over Clinical Trials: II Binary Outcomes. *Stat. Med.* **2002b,** *21,* 2145–2159.

Devereaux, P. J.; Choi, P. T. L.; Lacchetti, C.; Weaver, B.; Schunemann, H. J.; Haines, T.; Lavis, J. N.; Grant, B. J. B.; Haslam, D. R. S.; Bhandari, M.; Sullivan, T.; Cook, D. J.; Walter, S. D.; Meade, M.; Khan, H.; Bhatnagar, N.; Guyatt, G. H. A Systematic Review and Meta-analysis of Studies Comprising Mortality Rates of Private For-profit and Private Not-for-profit Hospitals. *Can. Med. Assoc. J.* **2002,** *166,* 1399–1406.

Dwyer, T.; Couper, D.; Walter, S. D. Sources of Heterogeneity in the Meta-analysis of Observational Studies: The Example of SIDS and Sleeping Position. *J. Clin. Epidemiol.* **2001,** *54,* 440–447.

Egger, M.; Smith, G. D.; Altman, D. G., Eds. *Systematic Reviews in Health Care—Meta-analysis in Context.* BMJ Publishing Group: London, 2001.

Egger, M.; Schneider, M.; Smith, G. D. Meta-analysis: Spurious Precision? Meta-analysis of Observational Studies. *Br. Med. J.* **1998,** *316,* 140–144.

Eysenck, H. J. The Effects of Psychotherapy. *Int. J. Psychiatry* **1965,** *1,* 97–178.

Ezzat, S.; Asa, S. L.; Couldwell, W. T.; Barr, C, E.; Dodge, W. E.; Vance, M. L.; McCutcheon, I. E. The Prevalence of Pituitary Adenomas: A systematic Review. *Cancer* **2004,** *101,* 613–619.

Devereaux, P. J.; Choi, P. T. L.; Lacchetti, C.; Weaver, B.; Schunemann, H. J.; Haines, T.; Lavis, J. N.; Grant, B. J. B.; Haslam, D. R. S.; Bhandari, M.; Sullivan, T.; Cook, D. J.; Walter, S. D.; Meade, M.; Khan, H.; Bhatnagar, N.; Guyatt, G. H. A Systematic Review and Meta-analysis of Studies Comprising Mortality Rates of Private for-Profit and Private not-for-Profit Hospitals. *Can. Med. Assoc. J.* **2002,** *166,* 1399–1406.

Dwyer, T.; Couper, D.; Walter, S. D. Sources of Heterogeneity in the Meta-analysis of Observational Studies: the Example of SIDS and Sleeping Position. *J. Clin. Epidemiol.* **2001,** *54,* 440–447.

Egger, M.; Schneider, M.; Smith, G. D. Meta-analysis: Spurious Precision? Meta-analysis of Observational Studies. *Br. Med. J.* **1998,** *316,* 140–144.

Fisher, L. D.; Belle, G. V. *Biostatistics: A Methodology for the Health Sciences.* New York: John Wiley and Sons Inc., 1993.

Glass, G. V. *Meta-Analysis at 25.* College of Education, Arizona State University: Arizona, 2000. http://glass.ed.asu.edu/gene/papers/meta25.html.

Glass, G. V. Primary, Secondary and Meta-analysis of Research. *Educ. Res.* **1976,** *10,* 3–8.

Gisbert, J. P.; Gracia-Buey, L.; Pajares, J. M.; Moreno-Otero, R. Prevalence of Hepatitis C Virus Infection in Porphyria Cutanea Tarda: Systematic Review and Meta-analysis. *J. Hepatol.* **2003,** *39,* 620–627.

Gupta, R.; Gupta, V. P. Meta-analysis of Coronary Heart Disease Prevalence in India. *Indian Heart J.* **1996,** *48,* 241–245.

Gupta, R. Meta-analysis of Prevalence of Hypertension in India. *Indian Heart J.* **1997,** *49,* 43–48.

Hall, J. A.; Roter, D. L. Do Patients Talk Differently to Male and Female Physicians? A Meta-analytic Review. *Pat. Educ. Counsel.* **2002,** *48,* 217–224.

Harrison, P. J.; Freemantle, N.; Geddes, J. R. Meta-analysis of Brain Weight in Schizophrenia. *Schizophr. Res.* **2003,** *64,* 25–34.

Hedges, L. V.; Olkin, I. Non-parametric Estimation of Effect Size in Meta-analysis. *Psychol. Bull.* **1984,** *96,* 573–580.

Herbert, T. B.; Cohen, S. Depression and Immunity: A Meta-Analytic Review. *Psychol. Bull.* **1993,** *113,* 472–486.

Kennedy, G. E.; Peersman, G.; Rutherford, G. W. International Collaboration in Conducting Systematic Reviews—The Cochrane Collaborative Review Group on HIV Infection and AIDS. *J. Acquir. Immune Defic.* **2002,** *30*(Supp. I), 56–61.

Klassen, T. P.; Jadad, A. R.; Moher, D. Guides for Reading and Interpreting Systematic Reviews. *Arch. Pediatr. Adolesc. Med.* **1998,** *152,* 700–7004.

Laird, N. M.; Ware, J. H. Random-Effects Models for Longitudinal Data. *Biometrics* **1982,** *38,* 963–974.

Lau, J.; Loannidis, J. P. A.; Schmid, C. H. Quantitative Synthesis in Systematic Reviews. *Ann. Intern. Med.* **1997**, *127*, 820–826.

Malhotra, S.; Karan, R. S.; Bhargava, V. K.; Pandhi, P.; Grover, A.; Sharma, Y. P.; Kumar, R. A Meta-analysis of Controlled Clinical Trials Comparing Low-molecular Weight Heparins with Unfractionated Heparin in Unstable Angina. *Indian Heart J.* **2001**, *53*, 197–202.

Moher, D.; Cook, D. J.; Eastwood, S.; Olkin, I.; Rennie, D.; Stroup, D. F. Improving the Quality of Reports of Meta-analyses of Randomized Controlled Trials—the QUOROM Statement. *Lancet* **1999**, *354*, 1896–1900.

Naylor, C. D. Meta-analysis and the Meta-epidemiology of Clinical Research. *Br. Med. J.* **1997**, *315*, 617–619.

Olsen, O.; Gotzsche, P. C. Cochrane Review on Screening for Breast Cancer with Mammography. *Lancet* **2001**, *358*, 1340–1342.

Pavia, M.; Bianco, A.; Pileggi, C.; Angelillo, I. F. Meta-analysis of Residential Exposure to Radon Gas and Lung Cancer. *Bull. World Health Organ.* **2003**, *81*, 732–738.

Reddy, M. V.; Hanji, M. B. Meta-analysis for Mental Health Care Research. *Indian J. Psychol. Med.* **2005**, *26*, 26–36.

Reynolds, K.; Lewis, L. B.; Nolen, J. D. L.; Kinney, G. L.; Sathya, B.; He, J. Alcohol Consumption and Risk of Stroke : A Meta-analysis. *J. Am. Med. Assoc.* **2003**, *289*, 579–587.

Satagopan, M. C.; Gupte, M. D.; Murthy, B. N.; Jabbar, S. Estimation of Cure, Relapse and Success Rates of Short-course Chemotherapy in the Treatment of Pulmonary Tuberculosis: A Meta-analysis. *Indian J. Tubercul.* **2002**, *49*, 87–96.

Sharma, P. K.; Malhotra, S.; Pandhi, P.; Kumar, N. Effect of Inhaled Steroids on Bone Mineral Density: A Meta-Analysis. *J. Clin. Pharmacol.* **2003**, *43*, 193–197.

Shann, F. Meta-analysis of Trials of Prophylactic Antibiotics for Children with Measles: Inadequate Evidence. *Br. Med. J.* **1997**, *314*, 334–337.

Shi, J. Q.; Copas, J. B. Meta-analysis for Trend Estimation. *Stat. Med.* **2004**, *23*, 3–19.

Sterne, J. A. C.; Egger, M.; Smith, G. D. Investigating and Dealing with Publication and Other Biases in Meta-analysis. *Br. Med. J.* **2001**, *323*, 101–105.

Stewart, W. F.; Simon, D.; Shechter, A.; Lipton, R. B. Population Variation in Migraine Prevalence: A Meta-Analysis. *J. Clin. Epidemiol.* **1995**, *48*, 269–280.

Stroup, D. F.; Berlin, J. A.; Morton, S. C.; Olkin, I.; Williamson, G. D.; Rennie, D.; Moher, D.; Beeker, B. J.; Sipe, T. A.; Thacker, S. B. Meta-analysis of Observational Studies in Epidemiology—A Proposal for Reporting. *J. Am. Med. Assoc.* **2000**, *283*, 2008–2012.

Strube, M. J.; Hartmann, D. P. Meta-Analysis: Techniques, Applications, and Functions. *J. Consult. Clin. Psychol.* **1983**, *51*, 14–27.

Tharyan, P. The Relevance of Meta-analysis, Systematic Reviews and the Cochrane Collaboration to Clinical Psychiatry. *Indian J. Psychiatry* **1998**, *40*, 135–148.

Todd, S. Incorporation of Sequential Trials into a Fixed Effects Meta-analysis. *Stat. Med.* **1997**, *16*, 2915–2925.

Walker, A. Meta-style and Expert Review. *Lancet* **1999**, *354*, 1834–1835.

Whitehead, A. A Prospectively Planned Cumulative Meta-Analysis Applied to a Series of Concurrent Clinical Trials. *Stat. Med.* **1997**, *16*, 2901–2913.

CHAPTER 2

PROTOCOL WRITING FOR META-ANALYSIS STUDY

CONTENTS

ABSTRACT

Defining a hypothesis and determining research questions must be specific and clear about what you really focus in it. Analytical meta-analysis is considered to invade the points of estimation, and exploratory meta-analysis is mainly focused on investigating potential source of heterogeneity and may reveal important effect modifiers. There are two basic approaches to combininge evidence across studies in meta-analysis. One approach involves testing the statistical significance of combined results of the collection of studies. The other approach involves estimating an average treatment effect. The sources of search for literature in meta-analysis include the published literature, unpublished literature, uncompleted research reports, and work in progress. Reliance on only published reports leads to publication bias—the bias resulting from the tendency to publish results that are statistically significant. Given a vast quantity of heterogeneous literature, suitable studies have to be selected for a meta-analysis. Meta-analysis is a two-stage process. In the first stage, the effect sizes are collected from each primary study. Methods of quality assessment provide a systematic approach to describe primary studies and explain heterogeneity. A formal approach to decide the ultimate inclusion criteria of a study may be undertaken by using panel of judges/experts. The internal validity of a study is the extent to which systematic error (bias) is minimized. The external validity is the extent to which the results of the study provide a correct basis for applicability to other circumstances. The first step in meta-analysis is to prepare a master sheet (data-points table). The first column in the master sheet consists of the list of selected studies according to their chronological order of publication. In more complex situations to understand heterogeneity and its sources, several graphs and diagrams such as Forest plot, Funnel plot, etc. have been established to use in meta-analysis. The methods used to pool end-points explain a weighted averages of the end-points in which the larger studies generally have more influence than the smaller ones. The methods are based on the assumptions such as fixed effects and random effects models. It will be advantageous to extend meta-analysis by applying several additional meta-analysis techniques such as sensitivity analysis techniques, influence meta-analysis technique, subgroup meta-analysis technique, and cumulative meta-analysis technique. In reporting meta-analysis, various implications of the results such as research implications, clinical implications, economic implications, and implications for policy making have to be specified.

2.1   RESEARCH PROBLEMS FOR META-ANALYSIS

Defining a hypothesis and determining research questions must be specific and clear about what you really focus it.

2.1.1   NEED FOR THE STUDY

Meta-analysis enables researchers to combine the results of many pieces of research on a topic to determine whether the findings hold generality. This is better than trying to assume that the finding of a suitable study has global meaning. The meta-analysis combines systematically the results of similar but independent studies whenever relevant studies on an interest have conflicting conclusions.

2.2   TYPE OF META-ANALYSIS

2.2.1   CLASSIFICATION OF LITERATURE

The literature of meta-analysis can be classified as the papers that deal with methodological and statistical issues, the papers actually carrying out meta-analysis and the review papers.

2.2.2   ANALYTICAL META-ANALYSIS AND EXPLORATORY META-ANALYSIS

Analytical meta-analysis is considered to invade the points of estimation, and exploratory meta-analysis mainly focused on investigating potential source of heterogeneity and may reveal important effect modifiers.

2.2.3   TWO APPROACHES

There are two basic approaches to combining evidence across studies in meta-analysis. One approach involves testing the statistical significance of combined results of the collection of studies. The other approach involves estimating an average treatment effect. A confidence interval or significant test is often used to determine whether the average effect is reliably different from some hypothetical value. Although the two approaches use different information from each study, the combined significance tests use p-values, and combined estimate procedures use measures of effect size, the methods are clearly related.

2.3   PLAN OF META-ANALYSIS STUDY

2.3.1   LOCATION OF STUDIES

The sources of search for literature in meta-analysis include the published literature, unpublished literature, uncompleted research reports, and work in progress. Reliance on only published reports leads to publication bias—the bias resulting from the tendency to publish results that are statistically significant (Reddy, 2014).

2.3.2   SELECTION OF STUDIES

Given a vast quantity of heterogeneous literature, suitable studies have to be selected for a meta-analysis. The inclusion and exclusion criteria relate to the quality and combinability of patients and outcome.

2.3.3   EFFECT SIZES OF PRIMARY STUDIES

Meta-analysis is a two-stage process. In the first stage the effect sizes (end points or summary statistics of studies) are collected from each primary study. The end points may be proportions, mean difference, odds ratio, Z-value, Cohen's d, etc. All the studies selected for a meta-analysis may provide different end points (data points). In such cases, a transformation to common end point is necessary. It is convenient to transform different statistics to the correlation coefficient r before proceeding with further analysis.

2.3.4   QUALITY ASSESSMENT OF SELECTED STUDIES

Methods of quality assessment provide a systematic approach to describe primary studies and explain heterogeneity. A formal approach to decide the ultimate inclusion criteria of a study may be undertaken by using panel of judges/experts. The quality assessment items include areas such as the report itself, the study, the patients, the study design, effect size, etc.

2.3.5   INTERNAL VALIDITY AND EXTERNAL VALIDITY

The internal validity of a study is the extent to which systematic error (bias) is minimized. The external validity is the extent to which the results of the study provide a correct basis for applicability to other circumstances.

2.4   STATISTICAL METHODS IN META-ANALYSIS

2.4.1   META-ANALYSIS MASTER SHEETS

The first step in meta-analysis is to prepare a master sheet (data points table). The first column in the master sheet consists of the list of selected studies according to their chronological order of publication. The last column of the table consists of their respective end points in order to notice the statistical heterogeneity. The information on relevant variables (quality assessment items) is ensured in the master sheet in order to note the clinical heterogeneity.

2.4.2   META-ANALYSIS PLOTS

In more complex situations to understand heterogeneity and its sources, several graphs and diagrams such as Forest plot, Funnel plot, etc. have been established to use in meta-analysis.

2.4.3 METHODS FOR POOLING EFFECT SIZES

The methods used to pool end points explain a weighted averages of the end points in which the larger studies generally have more influence than the smaller ones. The methods are based on the assumptions such as fixed effects and random effects models.

2.4.4 ADDITIONAL META-ANALYSIS TECHNIQUES

It will be advantageous to extend meta-analysis by applying several additional meta-analysis techniques such as sensitivity analysis techniques, influence meta-analysis technique, subgroup meta-analysis technique, and cumulative meta-analysis technique.

2.5 REPORTING THE RESULTS

In reporting meta-analysis, various implications of the results such as research implications, clinical implications, economic implications, and implications for policy making have to be specified.

KEYWORDS

- publication bias
- effect Size
- heterogeneity
- weighted averages
- implications

REFERENCE

Reddy, M. V. *Statistical Methods in Psychiatry Research*. Apple Academic Press Inc.: Canada, 2014.

CHAPTER 3

FORMULATION OF RESEARCH PROBLEM

CONTENTS

ABSTRACT

The first step in defining your research question is to decide what theoretical constructs to be used as explanatory and response variables. Once you have determined what effect you want to examine, you must determine the population in which you want to examine it. The first criterion you must have is that the studies need to measure both the explanatory and response variables defining your effect and provide an estimate of their relationship. You should expect that your list of inclusion and exclusion criteria will change during the course of your analysis. The first and most important decision in preparing a review is to determine its focus. This is best done by asking clearly framed questions. It is often helpful to consider the types of people that are of interest in two steps. First, define the diseases or conditions that are of interest. Second, identify the population and setting of interest. The next key component of a well-formulated question is to specify the interventions that are of interest. The third key component of a well-formulated question is the delineation of particular outcomes that are of interest. Determining the scope of a review question is a decision dependent upon multiple factors. Narrow questions may result in spurious or biased conclusions and may not be generalizable. As broad questions may be addressed by large sets of heterogeneous studies, the synthesis and interpretation of data may be particularly challenging.

3.1   DEFINITION OF RESEARCH QUESTION

The first step in defining your research question is to decide what theoretical constructs to be used as explanatory and response variables. There are several things to consider when selecting a hypothesis for meta-analysis:

(1) There should be a significant available literature, and it should be in a quantifiable form.
(2) The hypothesis should not require the analysis of an overwhelming number of studies.
(3) The topic should be interesting to others.
(4) There should be some specific knowledge to be gained from the analysis.

Some reasons to perform meta-analyses are to establish the presence of an effect, determine the magnitude of an effect, resolve differences in literature, and determine important moderators of an effect.

3.1.1 CHOICE OF EFFECT SIZE

If you decide to use the effect size d, you then need to precisely define what contrast you will use to calculate d. For a simple design, this will probably be (mean of experimental group—mean of control group). Defining the contrast also specifies the directionality of your effect size (i.e., the meaning of the sign). The directionality is automatically determined for the effect size r once you choose your constructs.

3.2 LIMITING THE PHENOMENON OF INTEREST

3.2.1 POPULATION TO BE STUDIED

Once you have determined what effect you want to examine, you must determine the population in which you want to examine it. If you are performing a meta-analytic summary you will often choose very practical boundaries for your population, such as the experiments reported in a specific paper. The populations for quantitative literature reviews, however, should be defined on a more abstract, theoretical level. In the latter case you, define a specific set of inclusion and exclusion criteria that studies must meet to be included in the analysis.

The goal of this stage is to define a population that is a reasonable target for synthesis. You want your limits narrow enough so that the included studies are all examining the same basic phenomenon, but broad enough so that there is something to be gained by the synthesis that could not easily be obtained by looking at an individual study.

When performing a meta-analytic summary you often limit your interest to establishing the presence of an effect and estimating its size. However, quantitative literature reviews should generally go beyond this and determine what study characteristics moderate the strength of the effect.

3.2.2 INCLUSION AND EXCLUSION CRITERIA

The first criterion you must have is that the studies need to measure both the explanatory and response variables defining your effect and provide an estimate of their relationship. Without this information there is nothing you can do with a study meta-analytically.

Each additional criterion that you use to define the population of your meta-analysis should be written down. Where possible, you should provide examples of studies that are included or excluded by the criterion to help clarify the rule.

You should expect that your list of inclusion and exclusion criteria will change during the course of your analysis. Your perception of the literature will be better informed as you become more involved in the synthesis, and you may discover that your initial criteria either cut out parts of the literature that you want to include, or else are not strict enough to exclude certain studies that you think are fundamentally different from those you wish to analyze. You should feel free to revise your criteria whenever you feel it is necessary, but if you do so after you've started coding you must remember recheck studies you've already completed.

It is a good practice to keep a list of the studies that turned up in your initial search but that you later decided to exclude from your analysis. You should also record exactly what criterion they failed to meet, so that if you later decide to relax a particular criterion you know exactly what studies you will need to reexamine, saving you from having to perform an entirely new literature search.

3.3 KEY COMPONENTS OF RESEARCH QUESTION

3.3.1 RATIONALE FOR RESEARCH QUESTION

The first and most important decision in preparing a review is to determine its focus (Light and Pillemer, 1984). This is best done by asking clearly framed questions. Such questions are essential for determining the structure of a review (Cooper and Hedges, 1994; Hedges and Olkin, 1984). Specifically, they will guide much of the review process including strategies for locating and selecting studies or data, for critically appraising their relevance and validity, and for analyzing variation among their results.

In addition to guiding the review process, a review's questions and objectives are used by readers in their initial assessments of relevance. The readers use the stated questions and objectives to judge whether the review is likely to be interesting and directly relevant to the issues they face.

3.3.2   TYPE OF PEOPLE

It is often helpful to consider the types of people that are of interest in two steps. First, define the diseases or conditions that are of interest. Explicit criteria sufficient for establishing the presence of the disease or condition should be developed. Second, identify the population and setting of interest. This involves deciding whether one is interested in a special population group determined on the basis of factors such as age, sex, race, educational status, or the presence of a particular condition such as angina or shortness of breath. One might also be interested in a particular setting on the basis of factors such as whether people are living in the community; are hospitalized, in nursing homes or chronic care institutions; or are outpatients. Any restrictions with respect to specific population characteristics or settings should be based on sound evidence.

3.3.3   TYPE OF COMPARISON

The next key component of a well-formulated question is to specify the interventions that are of interest. It is also important to define the interventions against which these will be compared, such as the types of control groups that are acceptable for the review.

3.3.4   TYPE OF OUTCOME

The third key component of a well-formulated question is the delineation of particular outcomes that are of interest.

3.3.5 TYPE OF STUDY DESIGN

Certain study designs are superior to others when answering particular questions. Randomized controlled trials are considered by many the *sine qua non* when addressing questions regarding therapeutic efficacy, whereas other study designs are appropriate for addressing other types of questions.

3.4 BROAD VERSUS NARROW QUESTION

Determining the scope of a review question is a decision dependent upon multiple factors including perspectives regarding a question's relevance and potential impact; supporting theoretical, biologic, and epidemiological information; the potential generalizability and validity of answers to the questions; and available resources.

3.4.1 NARROW QUESTIONS

Narrowly focused reviews may not be generalizable to multiple settings, populations, and formulations of an intervention. They can also result in spurious or biased conclusions in the same way that subgroup analyses sometimes do.

A narrow focus is at high risk of resulting in biased conclusions when the author is familiar with the literature in an area and narrows the inclusion criteria in such a way that one or more studies with results that are in conflict with the author's beliefs are excluded.

3.4.2 BROAD QUESTIONS

The validity of very broadly defined reviews may be criticized for mixing apples and oranges, particularly when there is good biologic or sociological evidence to suggest that various formulations of an intervention behave very differently or that various definitions of the condition of interest are associated with markedly different effects of the intervention.

Searches for data relevant to broad questions may be more time-consuming and more expensive than searches relevant to narrowly defined questions. As broad questions may be addressed by large sets of heterogeneous studies, the synthesis and interpretation of data may be particularly challenging.

3.5 CHANGING QUESTIONS

Although a certain fluidity and refinement of questions is to be expected in reviews as one gains a fuller understanding of the problem, it is important to guard against bias in modifying questions. Post-hoc questions are more susceptible to bias than those asked a priori, and data-driven questions can generate false conclusions based on spurious results. Any changes to the protocol that result from revising the question for the review should be documented.

KEYWORDS

- **response variables**
- **inclusion criteria**
- **exclusion criteria**
- **narrow questions**
- **broad questions**

REFERENCES

Cooper, H. M.; Hedges, L. V. *The Handbook of Research Synthesis*. Russell Sage Foundation: New York, 1994.

Hedges, L. V.; Olkin, I. Non-parametric Estimation of Effect Size in Meta-analysis. *Psychol. Bull.* **1984,** *96*, 573–580.

Light, R. J.; Pillemer, D. B. *Summing Up. The Science of Reviewing Research.* Harvard University Press: Cambridge, 1984.

CHAPTER 4

LOCATION AND SELECTION OF STUDIES

CONTENTS

ABSTRACT

The formulation of detailed objectives is at the heart of any research project. This should include the definition of study participants, interventions, outcome, and results. It is prerequisite to define any basic inclusion criteria to assess the methodological quality of comparable studies and to perform a thorough sensitivity analysis. It is useful to have two observers checking eligibility of candidate studies with disagreement being resolved by discussion or a third reviewer. Selection bias can also occur in making decision about which study to be included and which study to be excluded from meta-analysis. Ideally, all inclusion criteria are set before reviewing the studies identified. The source of search for literature in meta-analysis includes the published literature, unpublished literature, uncompleted research reports, and work in progress. These are also different methods of searching the literature. A backward search involves identifying a publication and then moving to earlier items in the citation. A forward search identifies a publication and then searches all items that later cite the publication. All the studies that fit within the criteria should be located. To provide an accurate estimate of an effect, it is important to find unpublished articles for analysis. Professional meetings are particularly good way to locate unpublished articles. It is good policy to write to the first author of each article that you decide to include in your analysis to see if they have any unpublished research relating to your topic. Research registers are actively maintained lists of studies centered on a common theme. Sometimes the number of studies that fit inside your boundaries is too large for you to analyze them all then choose a random sample of the studies for coding and analysis. The exclusion of gray literature from meta-analysis can lead to exaggerated estimates of intervention effectiveness. A standard strategy would include hand searching of selected journals and checking the references of all the relevant studies identified. Previous reviews, whether they include a meta-analysis or not, are often a fruitful place to look for relevant studies. Foreign studies should be included in the analysis unless you expect that cross-cultural differences would affect the results. The electronic searching, conference proceedings, nonindexed journals would not typically be included. Not all the reports retrieved are appropriate for inclusion in a meta-analysis. You would be well advised to make use database program to assist you in this task. Results of smaller studies can be expected to be more widely scattered around the average.

The personnel bias exists when some people have economic, social, or political agenda in order to favor their legislations.

4.1 SETTING INCLUSION AND EXCLUSION CRITERIA

4.1.1 OBJECTIVES

The formulation of detailed objectives is at the heart of any research project. This should include the definition of study participants, interventions, outcome, and results.

4.1.2 USE OF SENSITIVITY ANALYSIS

Inclusion and exclusion criteria in studies and eligibility criteria can be defined for the types of studies to be included. They relate to the quality and to the combinability of patients and outcome. It is prerequisite to define any basic inclusion criteria to assess the methodological quality of comparable studies and to perform a thorough sensitivity analysis.

4.1.3 DEGREE OF SUBJECTIVITY

Decision regarding the inclusion and exclusion of individual studies often involves some degree of subjectivity. It is therefore useful to have two observers checking eligibility of candidate studies with disagreement being resolved by discussion or a third reviewer (Egger et al., 2001). A standard record form is used for the purpose. Data extraction should carefully design, piloted, and revised if necessary.

4.1.4 REFINEMENT OF CRITERIA

Selection bias can also occur in making decision about which study to be included and which study to be excluded from meta-analysis. Ideally, all inclusion criteria are set before reviewing the studies identified. But it is not always practical, since investigator may already know the literature fairly well, and eligibility criteria certainly need refinement based on the available information. One might imagine that high quality meta-analysis, such as those that set eligibility criteria before beginning data collection, or those that make reviewers to authors and results, might have different findings than those that did not. This has to be tested by grouping reviewers using various methodologies and comparing average effect sizes obtained across reviewers of each type (Dickersin, 2002).

4.2   LITERATURE SEARCH

The source of search for literature in meta-analysis includes the published literature, unpublished literature, uncompleted research reports, and work in progress.

4.2.1   PUBLICATION BIAS

Reliance on only published reports leads to publication bias—the bias resulting from the tendency to publish results that are statistically significant. As a first step toward eliminating publication bias, the meta-analysis needs to obtain information from unpublished research. Registration of studies at the time they are established could eliminate the risk of publication bias. Contacts with colleagues, experts in the field, and other informed channels can also be important sources of information on unpublished and ongoing studies. Among the published studies, those with higher significant results are more likely to be published without delay, more likely to be cited, and more likely to be published more than once (Schwarzer et al., 2002; Stroup et al., 2001). Some studies may be impossible to retrieve and include in a meta-analysis, despite a thorough search of potential database. Publication bias is difficult to eliminate, but some statistical procedures may be helpful in detecting its presence. An inverted funnel plot is sometimes used to visually explore the possibility that publication bias is present.

4.2.2   LOCATION OF PUBLISHED REPORTS

The meta-analysis begins with search of regular bibliographic reports, cited indices and abstracted databases to provide information regarding published reports. These publications are retrieved, and the process is repeated again and again, manual search of databases requiring specification of a search statement, and a method of searching.

4.2.3   SEARCH PROCEDURE

These are also different methods of searching the literature. A backward search involves identifying a publication and then moving to earlier items in the citation. A forward search identifies a publication and then searches all items that later cite the publication.

4.3   EXHAUSTIVE SEARCH

All the studies that fit within the criteria should be located. When performing a meta-analysis commonly you will sometimes know at the state exactly what studies you want to locate. It needs to perform a detailed search to locate all the studies that have examined the effect of interest within the population defined. Search the literature to find possible candidates for analysis using fairly open guidelines. To locate all of the studies that truly meets the criteria, even if searches also included a large number of irrelevant studies.

4.3.1   UNPUBLISHED REPORTS

To provide an accurate estimate of an effect, it is important to find unpublished articles for your analysis. Many studies have shown that published articles typically favor significant findings over nonsignificant findings, which biases the findings of analysis based solely on published studies.

4.3.2   PROGRAMS FROM PROFESSIONAL MEETINGS

This is particularly good way to locate unpublished articles, since papers presented at conferences are typically subject to a less restrictive review than journal articles.

4.3.3   LETTERS TO ACTIVE RESEARCHERS

It is good policy to write to the first author of each article that you decide
to include in your analysis to see if they have any unpublished research
relating to your topic. When trying to locate people, you may want to
make use of academic department offices/department web pages, alumini
offices (to trace down the authors of dissertations), internet search engines,
and membership guides.

4.3.4   RESEARCH REGISTERS

Research registers are actively maintained lists of studies centered on a
common theme. Currently, there are very few research registers available for psychological research, but this may changes with the spread of
technology.

4.3.5   SAMPLING OF STUDIES

Sometimes the number of studies that fit inside your boundaries is too
large for you to analyze them all. In this case, you should still perform
an exhaustive search of the literature. Afterward, you choose a random
sample of the studies you found for coding and analysis.

4.4   GRAY LITERATURE

The exclusion of gray literature from meta-analysis can lead to exaggerated
estimates of intervention effectiveness. In general, meta-analysis should
attempt to identify, retrieve, and include all reports, gray and published
that meet predefined inclusion criteria.

4.4.1   HAND SEARCH

A standard strategy would include the search of a computerized database
including the identification of any review articles that might identify

studies not picked up in the searches, hand searching of selected journals and checking the references of all the relevant studies identified (Armitage and Colton, 1998). Elaborate searching of bibliographical database is probably the most common method used to identify studies. The hand search is not precise and complete, which forces readers to review thousands of irrelevant articles of abstract.

If you find that many of your articles are coming from a specific journal, then you should go back and read through the table of contents of that journal for all of the years that there was active research on your topic. You should make use of current contents, a journal containing a listing of the table of contents of other journals.

4.4.2   REFERENCE LIST OF REVIEW ARTICLES

Previous reviews, whether they include a meta-analysis or not, are often a fruitful place to look for relevant studies.

4.4.3   FOREIGN STUDIES

Foreign studies should be included in the analysis unless you expect that cross-cultural differences would affect the results and you lock enough foreign studies to test this difference.

4.4.4   ELECTRONIC SEARCH

The electronic searching, conference proceedings, nonindexed journals would not typically be included. Appropriate terms to index observational studies were introduced in the widely used bibliographic databases MEDLINE and EMBASE by the mid-1990s. The majority of journals indexed in MEDLINE are published in the United States, whereas EMBASE has better coverage of European journals. All studies identified in the retagging and hand searching projects have been included in the "The Cochrane Controlled Trials Registers," which is available in the Cochrane library on CD-ROM or online.

4.5   SELECTION OF STUDIES

Not all the reports retrieved are appropriate for inclusion in a meta-analysis. Some turn out to be having no data of any kind and some have collected data but report on the data so poorly that they are unusable. Some are border line cases where the meta-analyst is given enough data that good detective work allows him to obtain at least an approximate effect size estimate and significance level. Meta-analysis involves the summarization of data, not of an author's conclusion, so the previous statements are of little help to the meta-analyst. However, if the meta-analyst has the relevant means and standard deviations, he can compute the effect sizes. If in addition the sample sizes are given, the meta-analyst can also compute accurate p-values (Rosenthal, 1995).

4.5.1   MASTER CANDIDATE LIST

Performing a comprehensive search of the literature involves working with a huge amount of information. You would be well advised to make use database program to assist you in this task. For each study in the master candidate list you should record.

4.5.2   ANALYSIS OF FURTHER BIAS

It should be vital and integrated part in conducting a meta-analysis. A statistical test has been proposed for this purpose, namely, the rank correlation test (Begg and Mazumdar, 1994). For this test, the variance of the treatment effect in each single study is of certain importance. The smaller the studies are, the larger the random error is, and so too is the variation in results. Consequently, results of smaller studies can be expected to be more widely scattered around the average.

4.2.3   PERSONNEL BIAS

The personnel bias exists when some people have economic, social, or political agenda in order to favor their legislations. These persons want to incorporate larger favorable data sets in the meta-analysis.

KEYWORDS

- **selection bias**
- **backward search**
- **forward search**
- **publication bias**

REFERENCES

Armitage, P.; Colton, T. *Encyclopedia of Biostatistics.* John Wiley & Sons: New York, 1998; Vol 4.

Begg, G. B.; Mazumdar, M. Operating Characteristics of a Rank Correlation Test for Publication Bias. *Biometrics* **1994**, *50*, 1088–1101.

Egger, M.; Smith, G. D.; Altman, D. G., Eds. Systematic Reviews in Health Care—Meta-analysis in Context. BMJ Publishing Group: London, 2001.

Dickersin, K. Systematic Reviews in Epidemiology: Why are We So Far Behind? *Int. J. Epidemiol.* **2002**, *31*, 6–12.

Rosenthal, R. Writing Meta-analytic Reviews. *Psychol. Bull.* **1995**, *118*, 183–192.

Schwarzer, G.; Antes, G.; Schumacher, M. Inflation of Type I Error Rate in Two Statistical Tests for the Detection of Publication Bias in Meta-Analyses with Binary Outcomes. *Stat. Med.* **2002**, *21*, 2465–2477.

Stroup, D. F.; Thacker, S. B.; Olson, C. M.; Glass, R. M.; Hutwagner, L. Characteristics of Meta-analyses Related to Acceptance for Publication in a Medical Journal. *J. Clin. Epidemiol.* **2001**, *54*, 655–660.

CHAPTER 5

QUALITY ASSESSMENT OF SELECTED STUDIES

CONTENTS

ABSTRACT

Given a vast quantity of heterogeneous literature, the type of items that should be collected should include the report of the study, the study itself, the patients, the research design, the effect size, and the methodological quality. In the context of a systematic review, the validity of a study is the extent to which its design and conduct are likely to prevent systematic errors or bias. The internal validity of a study is the extent to which systematic error (bias) is minimized. The external validity is the extent to which the results of the study provide a correct basis for applicability to other circumstances. There are many checklists and scales available to be used as evaluation tools. Quality assessment of individual studies that are summarized in meta-analysis is necessary to limit bias in conducting the systematic review, gain insight into potential comparisons, and guide interpretation of findings. Various sources of bias are selection bias, performance bias, attrition bias, detection bias, and time-lag bias. There are two major difficulties with assessing the validity of studies. The first is inadequate reporting of trials. The second is evidence of a relationship between parameters thought to measure validity and actual study outcomes.

5.1   QUALITY ASSESSMENT ITEMS

Given a vast quantity of heterogeneous literature, the type of items that should be collected should include the characteristics on

(1) The report of the study (such as author, year, and source)
(2) The study itself (scope, population)
(3) The patients (demographic factors, clinical features)
(4) The research design (experimental or observational, treatment-assignment mechanism or sampling mechanism, attrition rate or nonresponse rate)
(5) The effect size (sample size, nature of outcome, estimates, and standard error)
(6) The methodological quality (the internal validity and the external validity).

5.2 VALIDITY

In the context of a systematic review, the validity of a study is the extent to which its design and conduct are likely to prevent systematic errors, or bias (Moher et al., 1999). An important issue that should not be confused with validity is precision. Precision is a measure of the likelihood of chance effects leading to random errors. It is reflected in the confidence interval around the estimate of effect from each study and the weight given to the results of each study when an overall estimate of effect or weighted average is derived. More precise results are given more weight.

Variation in validity can explain variation in the results of the studies included in a systematic review. More rigorous studies may be more likely to yield results that are closer to the "truth." Quantitative analysis of results from studies of variable validity can result in "false-positive" conclusions. It is important to systematically complete critical appraisal of all studies in a review even if there is no variability in either the validity or results of the included studies.

5.2.1 INTERNAL VALIDITY AND EXTERNAL VALIDITY

The internal validity of a study is the extent to which systematic error (bias) is minimized. Such biases are the selection bias, performance bias, detection bias, and attrition bias. The external validity is the extent to which the results of the study provide a correct basis for applicability to other circumstances. Thus, the internal validity is a prerequisite for external validity (Jadad et al., 1998).

The readers blinded to the author, source, results, and discussion of the primary studies make the assessment of the quality of the study. The end result is a percentage score that may be incorporated into a sensitivity analysis at the analytical stage. The method of quality assessment provides a systematic approach to describe primary studies and explain heterogeneity. A formal approach to decide the ultimate inclusion status of a study may be undertaken using a panel of judges/experts.

Inadequate quality of studies may distort the results from meta-analysis. The use of summary scores from quality scales is problematic. Results depend on the choice of the scale, and the interpretation of findings is difficult. It is therefore preferable to examine the influence of individual components of methodological quality. Randomization versus nonrandomization, blinding of outcome assessment, and handling of patients attrition in the analysis should be assessed.

5.3   CHECKLIST AND SCALES

There are many checklists and scales available to be used as evaluation tools, but most are missing important evidence-based items when compared against the quality of reporting of meta-analyses (QUOROM) checklist, a gold standard. Beyond the generic features of study design and conduct, general quality scoring system may have to be supplemented or replaced with more problem-specific quality items for each particular meta-analysis. Empirical investigations have shown that studies of worse quality may overestimate prevalence rates (Hanji and Reddy, 2005a,b). Juni et al. (2001) have assessed the quality of controlled clinical trials during their systematic review in health care.

5.3.1   ASSESSMENT OF STUDY QUALITY

Quality assessment of individual studies that are summarized in meta-analysis is necessary to limit bias in conducting the systematic review, gain insight into potential comparisons, and guide interpretation of findings. Factors that warrant assessment are those related to applicability of findings, validity of individual studies, and certain design characteristics that affect interpretation of results.

Interpretation of results is dependent upon the validity of the included studies and other characteristics.

5.4   VARIOUS SOURCES OF BIAS

5.4.1   SELECTION BIAS

One of the most important factors that may lead to bias and distort treatment comparisons is that which can result from the way that comparison groups are assembled. Using an appropriate method for preventing foreknowledge of treatment assignment is crucially important in trial design.

5.4.2   PERFORMANCE BIAS

Performance bias refers to systematic differences in the care provided to the participants in the comparison groups other than the intervention under investigation. To protect against unintended differences in care and placebo effects, those providing and receiving care can be "blinded" so that they do not know the group to which the recipients of care have been allocated.

5.4.3   ATTRITION BIAS

Attrition bias refers to systematic differences between the comparison groups in the loss of participants from the study. It has been called exclusion bias. It is called attrition bias here to prevent confusion with pre-allocation exclusion and inclusion criteria for enrolling participants.

5.4.4   DETECTION BIAS

Detection bias refers to systematic differences between the comparison groups in outcome assessment. Trials that blind the people who will assess outcomes to the intervention allocation should logically be less likely to be biased than trials that do not.

5.4.5   TIME-LAG BIAS

Studies continued to appear in print media many years after approval by the ethics committee. The time lag was attributable to difference in the time from completion to publication. These findings indicate that time-lag bias may be introduced in meta-analysis even in situations when most or all trials will eventually be published. Trials with positive results will dominate the literature and introduce bias for several years until negative results but equally important results finally appear.

5.5   LIMITATIONS OF QUALITY ASSESSMENT

There are two major difficulties with assessing the validity of studies. The first is inadequate reporting of trials (SORT 1994, Schulz 1994, WGRR 1994, Begg 1996). It is possible to assume if something was not reported it was not done. However, this is not necessarily correct. Authors should attempt to obtain additional data from investigators as necessary. The second limitation, which in part evidence of a relationship between parameters thought to measure validity and actual study outcomes.

KEYWORDS

- **checklists**
- **scales**
- **quality assessment**
- **sources of bias**
- **validity of studies**

REFERENCES

Hanji, M. B.; Reddy, M. V. Meta-analytical Approach to Estimation of Pattern of Prevalence of Mental Retardation in India. *Indian J. Psychol. Med.* **2005a,** *27*(1), 103–111.

Hanji, M. B.; Reddy, M. V. Pattern of Prevalence of Affective Disorders in India: A Meta-analysis. *Indian J. Psychol. Med.* **2005b,** *27*(1), 155–164.

Jadad, A. R.; Moher, D.; Klassen, T. P. Guides for Reading and Interpreting Systematic Reviews—II. How did the Authors Find the Studies and Assess their Quality. *Arch. Pediatr. Adolesc. Med.* **1998,** *152*, 812–817.

Juni, P.; Altman, D. G.; Egger, M. Systematic Reviews in Health Care: Assessing the Quality of Controlled Clinical Trials. *Br. Med. J.* **2001,** *323*, 42–46.

Moher, D.; Cook, D. J.; Eastwood, S.; Olkin, I.; Rennie, D.; Stroup, D. F. Improving the Quality of Reports of Meta-analyses of Randomized Controlled Trials—The QUOROM Statement. *Lancet* **1999,** *354*, 1896–1900.

CHAPTER 6

EFFECT SIZES OF PRIMARY STUDIES

CONTENTS

ABSTRACT

An effect size is a unit-free quantitative measure of the strength of a phenomenon and independent of sample sizes. While combining effect sizes in meta-analysis, the standard error of effect-size plays an important role. As in any statistical setting, effect sizes are estimated with sampling error and may be biased unless the effect size estimator that is used is appropriate for the manner in which the data were sampled and the manner in which the measurements were made. The term effect size can refer to a standardized measure of effect (such as r, Cohen's d, and odds ratio), or to an unstandardized measure (the raw difference between group means). The first step to meta-analyzing a sample of studies is to describe the general distribution of effect sizes. An important moderator that has a strong influence on effect size may be considering separately for its descriptive analyses on each subpopulation.

One should always put effort into interpreting the observed effect sizes. There are three important effect sizes, namely, the Glass c, Cohen's d and Hedges g used in standardized difference in means for studies with different scales. All studies found in the literature may not provide the appropriate effect sizes. Instead, some may report Φ-value, P-value, χ^2-value, or any other statistics. In such cases, a transformation to a common endpoint is necessary. The sampling distribution of a correlation coefficient is somewhat skewed, especially if the population correlation is large. It is therefore conventional in meta-analysis to convert correlations to z scores using Fisher's r-to-z transformation. In between-effects test statistic, if you have access to the means and standard deviations of your two groups, you can calculate g from the definitional formula. For a study with binary outcome, the summary statistics include proportion of events in case of open trials, and odds ratio, risk ratio, and risk difference in case of controlled studies.

6.1   FEATURES OF EFFECT SIZE

6.1.1   STANDARD ERROR

An effect size is a unit-free quantitative measure of the strength of a phenomenon and independent of sample sizes. This facilitates measurement of an effect across the groups. While combining effect sizes in meta-analysis, the standard error (SE) of effect size plays an important role.

6.1.2   PARAMETER

The term "effect size" can refer to the value of a statistic calculated from a sample of data, the value of a parameter of a hypothetical statistical population, or to the equation that operationalizes how statistics or parameters lead to the effect-size value.

6.1.3   SAMPLING ERROR

As in any statistical setting, effect sizes are estimated with sampling error and may be biased unless the effect size estimator that is used is appropriate for the manner in which the data were sampled and the manner in which the measurements were made.

6.1.4   STANDARDIZED EFFECT SIZE

The term "effect size" can refer to a standardized measure of effect (such as r, Cohen's d, and odds ratio [OR]) or to an unstandardized measure (the raw difference between group means). Standardized effect size measures are typically used when the metrics of variables being studied do not have intrinsic meaning (a score on a personality test on an arbitrary scale), when results from multiple studies are being combined, when some or all of the studies use different scales, or when it is desired to convey the size of an effect relative to the variability in the population. In meta-analyses, standardized effect sizes are used as a common measure that can be calculated for different studies and then combined into an overall summary.

The methods for analyzing effect sizes are the same no matter what exact definition (mean difference, correlation) we decide to use.

6.1.5   REPORTING EFFECT SIZE

Reporting effect sizes is considered a good practice when presenting empirical research findings in many fields. The reporting of effect sizes facilitates the interpretation of the substantive, as opposed to the statistical significance of a research result. Effect sizes are particularly prominent in social and medical research. Relative and absolute measures of effect size convey different information and can be used complementarily.

6.1.6 EFFECT SIZE DISTRIBUTION

The first step to meta-analyzing a sample of studies is to describe the general distribution of effect sizes. A good way to describe a distribution is to report

1. Center of the distribution
2. General shape of the distribution
3. Significant deviations from the general shape

6.1.7 MODERATORS

One should closely examine any outlying effect sizes to ensure that they are truly part of the population one wishes to analyze. An important moderator that has a strong influence on effect size may be considering separately for its descriptive analyses on each subpopulation.

6.1.8 RAKING THE EFFECT SIZE

One should always put effort into interpreting the observed effect sizes. This will help in an intuitive understanding of the results. If other meta-analyses have been performed in related topic areas, one can report the mean size of those effects to provide context for the interpretation of study effect. If no other meta-analyses have been performed on related topics, one can compare the observed effect size to Cohen's (1992) guidelines as given below:

Size of effect	d	r
Small	0.2	0.1
Medium	0.5	0.3
Large	0.8	0.5

6.2 CONTINUOUS DATA EFFECT SIZE

6.2.1 MEAN DIFFERENCE EFFECT

For a study with continuous outcome, the summary statistics are the mean response in case of open studies and standardized difference in means for controlled studies. For a controlled study with continuous outcome, the results may be presented in a table as shown below.

Groups	Mean response	Standard deviation	Group size
Intervention	m_e	SD_e	n_e
Control	m_c	SD_c	n_c
Total			n

The pooled estimate of standard deviation of the two groups is computed as

$$s = \sqrt{\frac{(n_e - 1)SD_e^2 + (n_c - 1)SD_c^2}{n_e + n_c - 2}}$$

In a study with same scale, the difference in means and its SEs are given by

$$MD = (m_e - m_c)$$

$$SE(MD) = \sqrt{\frac{SD_e^2}{n_e} + \frac{SD_c^2}{n_c}}$$

There are three important effect sizes, namely, the Glass "c," Cohen's "d" and Hedges "g" used in standardized difference in means for studies with different scales.

6.2.2 GLASS "c"

It is given by

$$c = \frac{(m_e - m_c)}{SD_c}$$

$$SE(c) = \sqrt{\frac{n}{n_e n_c} + \frac{c^2}{2(n_c - 1)}}$$

6.2.3 COHEN'S d

It is given by

$$d = \frac{(m_e - m_c)}{s}$$

$$SE(d) = \sqrt{\frac{n}{n_e n_c} + \frac{d^2}{2(n - 2)}}$$

6.2.4 HEDGES "g"

It is given by

$$g = \frac{(m_e - m_c)}{s}\left(1 - \frac{3}{4n - 9}\right)\left(1 - \frac{3}{4n - 9}\right)$$

$$SE(g) = \sqrt{\frac{n}{n_e n_c} + \frac{g^2}{2(n - 3.94)}}$$

6.3   TRANSFORMATION TO EFFECT SIZE "r"

All studies found in the literature may not provide the appropriate effect sizes. Instead, some may report Φ-value, p-value, χ^2-value, or any other statistics. In such cases, a transformation to a common endpoint is necessary. It is convenient to transform different statistics to the correlation coefficient "r" before proceeding with further analysis. The r is unit free and its appropriate SE is given by

$$SE(r) = \sqrt{\frac{1}{n-3}}$$

The following is a list of formulas to transform several statistics to r:

(1) $r = \Phi$ in 2×2 contingency table.

(2) $r = CC$ in $r \times k$ contingency table.

(3) $r = r_{pb} \times 1.25$ when r_{pb} is given.

(4) $r = z/\sqrt{n}$ when z is given or obtained from p-value.

(5) $r = \sqrt{t^2/(t^2 + df)}$ when t-value is given or obtained as $t = \sqrt{F}$.

(6) $r = SS_B/(SS_B + SS_W)$ in ANOVA table.

(7) $r = \sqrt{d^2/(d^2 + 4)}$ when Cohen's d is given.

(8) $r = \sqrt{n_1 n_2 g^2/(n_1 n_2 g^2 + n_1 n_2 df)}$ when Hedges "g" is given.

Matt (1989) has analyzed rules for selecting effect sizes and showed that average effect estimates also varied with the rules. Such results indicate that the average effect estimates derived from meta-analysis may depend heavily on judgmental factors that enter into how effect sizes are selected within each of the individual studies considered relevant to a meta-analysis. Rosenthal and Rubin (1986) have presented a general set of meta-analytical procedures for combining studies with multiple effect sizes based on multiple-dependent variables.

6.3.1   FURTHER TRANSFORMATIONS

The sampling distribution of a correlation coefficient is somewhat skewed, especially if the population correlation is large. It is therefore conventional in meta-analysis to convert correlations to z scores using Fisher's r-to-z transformation

$$Z_r = \frac{1}{2}\ln\left(\frac{1+r}{1-r}\right)$$

where $\ln(x)$ is the natural logarithm function. All meta-analytic calculations are then performed using the transformed values.

There are several statistics that can be calculated from dichotomous variables that are related to correlation:

r_b **(biserial r):** This measures the relationship between two continuous variables when one of them is artificially dichotomized. It is an acceptable estimate of the underlying correlation between the variables.

r_{tet} **(tetrachoric r):** This measures the relationship between two continuous variables when both of them are artificially dichotomized. It is also an acceptable estimate underlying correlation.

r_{pb} **(point-biserial r):** This measures the relationship between a truly dichotomous variable and a continuous variable. It is actually a poor estimate of r, so we usually transform r_{pb} to r_b using the equation:

$$r_b = \frac{r_{pb}\sqrt{n_e n_c}}{\left|z^*\right|(n_e + n_c)}$$

where z^* is the point on the normal distribution with a p-value of $n_e/(n_{e+}n_c)$

$r_{\phi\phi}$ **(phi coefficient):** This measures the relationship between two truly dichotomous variables. This actually is an r.

If you have a t statistic you can calculate r_{pb} using the formula:

$$r_{pb} = \sqrt{\frac{t^2}{t^2 + n_e + n_c - 2}}$$

You can then transform r_{pb} into r_b using the above equation to get an estimate of r.

If you have a 1 df F statistic, you can calculate r_{pb} using the formula:

$$r_{pb} = \sqrt{\frac{F}{F + n_e + n_c - 2}}$$

You can then transform r_{pb} into r_b using equation starting with r_b to get an estimate of r.

If you have a 2 × 2 table for the response frequencies within two truly dichotomous variables, you can calculate r_ϕ from a chi-square test using the equation:

$$r_{\hat{\phi}} = \sqrt{\frac{\chi^2}{n}}$$

If you have a Mann–Whitney U (a rank-order statistic) you can calculate r_{pb} using the formula:

$$r_{pb} = 1 - \frac{2U}{n_e n_c}$$

where n_e and n_c are the sample sizes of your two groups. To get an estimate of r you can then transform r_{pb} to r_b using equation starting with r_b

You can calculate r from g using the equation:

$$r = \sqrt{\frac{g^2 n_e n_c}{g^2 n_e n_c + (n_e + n_c)(n_e + n_c - 2)}}$$

You should always report r to 4 decimal places.

Once the primary studies have been collected and coded, the meta-analyst needs to identify a summary measures common to all studies. Often the meta-analyst has little control over the choice of the summary measure because most of the decision is dictated by what was employed in the primary studies. In many settings, however, different summary measures will be reported across the primary studies. It now becomes the job of the analyst to create a summary that is comparable across all the studies (Normand, 1999). The summary statistics mainly depend upon the type of study such as experimental, observational, prognostic, and diagnostic in nature.

6.4   TRANSFORMATION TO EFFECT SIZE "g"

In between-effects test statistic, if you have access to the means and standard deviations of your two groups, you can calculate g from the definitional formula:

$$g = \frac{\tilde{Y}_e - \tilde{Y}_c}{S_p}$$

where $\tilde{Y}_e$ is the mean of the experimental group, $\tilde{Y}_c$ is the mean of the control group, and S_p is the pooled sample standard deviation.

If you have a between-subjects t statistic comparing the experimental and control groups,

$$g = t\sqrt{\frac{n_e + n_c}{n_e n_c}}.$$

From the same logic, if you have a between-subjects z-score comparing the experimental and control groups,

$$g = z\sqrt{\frac{n_e + n_c}{n_e n_c}}.$$

When you have the same number of subjects in the experimental and control group, this equation resolves to

$$g = \frac{2z}{\sqrt{2n}}$$

If you have a 1 numerator df F statistic comparing the experimental and control groups (we never directly calculate g from F statistics with more than 1 numerator df),

$$g = \sqrt{\frac{F\left(n_e + n_c\right)}{n_e n_c}}$$

If you have the same number of subjects in the experimental and control groups, this equation resolves to

$$g = \sqrt{\frac{2F}{n}}$$

The general formula for g in within-subject designs is

$$g = \frac{\tilde{Y}_e - \tilde{Y}_c}{S_{e-c}}$$

You can calculate the effect size from within-subjects test statistics using the formulas:

$$g = \frac{t}{\sqrt{n}}$$

and

$$g = \frac{z}{\sqrt{n}}$$

If you only have a p-value from a test statistic, you can calculate g if you know the direction of the finding. The basic procedure is to determine the test statistic corresponding to the p-value in a distribution table, and then calculate g from the test statistic.

You can get inverse probability distributions from a number of statistical software packages, including SAS. Even some hand-held calculators will provide the inverse distribution of the simpler statistics.

While an exact p-value allows an excellent estimate of a test statistic (and therefore g), a significance level (e.g., $p < 0.05$) gives a poorer estimate. You would treat significance levels as if it were an exact p-value in your calculations (e.g., treat $p < 0.05$ as $p = 0.05$).

6.4.1 CALCULATING G FROM DICHOTOMOUS-DEPENDENT VARIABLES

In case of dichotomous-dependent variables, that is in a 2 × 2 table, then $\chi^2 = z_2$. You may therefore get an unbiased estimate of the effect size from the equation:

$$g = \chi^2 \sqrt{\frac{n_e + n_c}{n_e n_c}}$$

When you have the same number of subjects in the experimental and control group, this equation resolves to

$$g = \sqrt{\frac{\chi^2}{n}}$$

You can alternatively calculate the phi-coefficient using the equation:

$$r_\delta = \sqrt{\frac{\chi^2}{n}}$$

If one or both levels have more than two levels, you can calculate

$$P = \sqrt{\frac{\chi^2}{n + \chi^2}}$$

6.4.2 MISCELLANEOUS

To calculate g from r you use the formula:

$$g = \frac{2r}{\sqrt{1 - r^2}}$$

To calculate g from nonparametric statistics you can find the p-value associated with the test and solve it for t.

You should always report g and d statistics to four decimal places.

6.5 BINARY DATA

For a study with binary outcome, the summary statistics include proportion of events in case of open trials and OR, risk ratio (RR), and risk difference (RD) in case of controlled studies. For a controlled study with a binary outcome, the results can be presented in a 2 × 2 table as shown below.

Group	Event	No event	Total
Intervention	a	b	n_1
Control	c	d	n_2

6.5.1 ODDS RATIO

The OR is given by

$$OR = \frac{(a/b)}{(c/d)}$$

The SE of its logarithm is given by

$$SE(\ln OR) = \sqrt{\frac{1}{a} + \frac{1}{b} + \frac{1}{c} + \frac{1}{d}}$$

6.5.2 RISK RATIO

The RR is given by

$$RR = \frac{(a/n_1)}{(c/n_2)}$$

The SE of its logarithm is given by

$$SE(\ln RR) = \sqrt{\frac{1}{a} + \frac{1}{c} - \frac{1}{n_1} - \frac{1}{n_2}}$$

6.5.3   RISK DIFFERENCE

The RD is given by

$$RD = \left(a/n_1\right) - \left(c/n_2\right)$$

The SE is given by

$$SE(RD) = \sqrt{\frac{ab}{n_1^3} + \frac{cd}{n_2^3}}$$

KEYWORDS

- effect size
- sampling error
- transformation
- odds ratio
- risk ratio
- risk difference

REFERENCES

Matt, G. E. Decision Rules for Selecting Effect Sizes in Meta-analysis: A Review and Reanalysis of Psychotherapy Outcome Studies. *Psychol. Bull.* **1989,** *105*, 106–115.

Normand, S. T. Meta-analysis: Formulating, Evaluating, Combining, and Reporting. *Stat. Med.* **1999,** *18*, 321–359.

Rosenthal, R.; Rubin, D. B. Meta-analytic Procedures for Combining Studies with Multiple Effect Sizes. *Psychol. Bull.* **1986,** *99*, 400–406.

CHAPTER 7

PREPARATION OF META-ANALYSIS MASTER SHEET

CONTENTS

ABSTRACT

The effect sizes measure along with identification particulars for each study has to be presented in the form of a table called meta-analysis master sheet. The important factors influencing the outcome of the studies have to be presented in order to view the clinical heterogeneity between studies. Such characteristics are known as moderators. The studies have to be listed according to chronological order of their publication. A unique number of every study called identification number has to be included in the meta-analysis master sheet.

Sometimes there is correlation between the study characteristics with the effect size. In such cases, these characteristics are called moderators. We precisely specify exactly how each moderator will be coded. Once the master sheet is prepared, the meta-analyst should analyze both the moderators, effect sizes with respect to their characteristic distributions such as the central tendency, degree of scatter, and the shape of the distributions.

7.1   META-ANALYSIS MASTER SHEET

The effect sizes measure along with identification particulars for each study has to be presented in the form of a table called meta-analysis master sheet. The important factors influencing the outcome of the studies have to be presented in order to view the clinical heterogeneity between studies. Such characteristics are known as moderators. The studies have to be listed according to chronological order of their publication (Reddy, 2014).

7.1.1   IDENTIFICATION PARTICULARS

A unique number of every study called identification number has to be included in the meta-analysis master sheet. A short or long reference of each study is also preferable.

7.1.2   IMPORTANT CHARACTERISTICS AND MODERATORS

A particular researcher's study quality has to be included. We can use these either as moderating variables or as basis for exclusion. One good way to code quality is to read through a list of validity threats Cook and Campbell, 1979) and consider whether each might have influenced studies in the analysis.

It is needed to record information about the overall design of the study such as assignment of subjects, experimental design, manipulation of codes, correlation definition and calculation procedures, and its number of subject that was measured to calculate the correlation. When using effect size "d," contrast definition, dependent measure, calculation method source of means, and source of standard errors have to be recorded.

7.2 MODERATOR ANALYSIS

Sometimes there is correlation between the study characteristics with the effect size. In such cases, these characteristics are called moderators. The moderating variables include particulars on major methodological variation, theoretical constructs, and basic study characteristics.

The test of moderating variables depends a great deal on the distribution of that variable in your sample. If most of your studies have the same value on a variable, then a test on that variable will not likely be informative. We should therefore try to select moderators that possess variability across sample of studies. The variables that we decide to code as moderators may also change as we learn more about the literature.

7.3 CODING MODERATING VARIABLES

We precisely specify exactly how each moderator will be coded. Sometimes the values that you assign to a moderator variable are fairly obvious, such as the year of publication. Other times, however, the assignment requires a greater degree of inference, such as when judging study quality. You should determine specific rules regarding how to code such "high-inference" moderators. If you have any high-inference coding that might be influenced by coder biases you should either come up with a set of low-inference codes that will provide the same information, or have the coding performed by individuals not working on the meta-analysis.

We make sure to code all the important characteristics that you think might moderate your effect. There is a tradeoff, however, in that analyzing a large number of moderators does increase the chance of you finding significant findings where they don't actually exist. If you have many moderators you might consider performing a multiple regression analysis including all of the significant predictors of effect size. The results of the multiple regressions automatically take the total number of moderators into account.

7.4 ANALYSIS OF DATA POINTS IN META-ANALYSIS

Once the master sheet is prepared, the meta-analysis should analyze both the moderators and effect sizes with respect to their characteristic distributions such as the central tendency, degree of scatter, and the shape of the distributions. The relationship between the moderators and the effect sizes have to be determined using either a chi-square test or correlation coefficient test. It is also necessary to check either the relationship is causal or it may be the action of a third variable.

7.5 NUMERICAL EXAMPLES FROM EPILEPSY STUDY

7.5.1 MASTER SHEET

The master sheet for 47 selected studies for prevalence studies of Epilepsy in India is presented in Table 7.1.

TABLE 7.1   Studies Selected for Meta-analysis in Prevalence of Epilepsy in India.

Sl. No.	Chief investigator	Year of report	State/UT	Locality	No. of families	Average family size	No. of persons	No. of cases	Preva-lence rate
1	Surya	1964	Pondicherry	Semi-urban	532	5.1	2731	6	2.20
2	Sethi	1967	UP	Urban	300	5.8	1733	2	1.15
3	Gopinath	1968	Karnataka	Rural	82	5.2	423	1	2.36
4	Dube	1970	UP	Mixed	6038	4.9	29,468	94	3.19
5	Elnagar	1971	WB	Rural	184	7.5	1383	6	4.34
6	Sethi	1972	UP	Rural	500	5.4	2691	6	2.23
7	Verghese	1973	TN	Semi-urban	539	5.4	2904	8	2.76

TABLE 7.1 *(Continued)*

Sl. No.	Chief investigator	Year of report	State/UT	Locality	No. of families	Average family size	No. of persons	No. of cases	Preva-lence rate
8	Sethi	1974	UP	Urban	850	5.3	4481	16	3.57
9	Nandi	1975	WB	Rural	177	6.0	1060	11	10.38
10	Carstairs	1976	Karnataka	Rural	344	6.2	2126	7	5.68
11	Nandi	1976	WB	Rural	177	6.1	1078	11	10.20
12	Nandi	1977	WB	Rural	590	4.9	2918	9	3.08
13	Agarwal	1978	Gujarat	Urban	200	5.1	1019	5	4.91
14	Nandi	1978a	WB	Rural	477	4.7	2230	15	6.73
15	Nandi	1978b	WB	Rural	450	5.0	2250	15	6.67
16	Nandi	1979	WB	Rural	609	6.1	3718	17	4.57
17	Nandi	1980a	WB	Rural	815	5.0	4053	13	3.21
18	Nandi	1980b	WB	Mixed	404	4.6	1862	10	5.37
19	Isaac	1980	Karnataka	Rural	733	5.7	4203	13	3.09
20	Bhide	1982	Karnataka	Rural	–	–	3135	15	4.79
21	Sen	1984	WB	Urban	337	6.4	2168	7	3.23
22	Mehta	1985	TN	Rural	1195	5.0	5941	44	7.41
23	Banerjee	1986	WB	Urban	205	3.8	771	2	2.59
24	Mathai	1986	TN	Mixed	–	–	45,778	411	8.98
25	Sachdeva	1986	Punjab	Rural	376	5.3	1989	5	2.51
26	Gourie-Devi	1987	Karnataka	Mixed	10,139	5.7	57,660	267	4.63
27	ICMR	1987a	Karnataka	Rural	–	–	35,548	278	7.82
28	ICMR	1987b	Gujarat	Rural	–	–	39,655	51	1.29
29	ICMR	1987c	WB	Rural	–	–	34,582	59	1.71
30	ICMR	1987d	Punjab	Rural	–	–	36,595	116	3.17
31	Bharucha	1988	Maharashtra	Urban	4537	3.1	14,010	66	4.71
32	Koul	1988	Jammu Kashmir	Rural	–	–	63,645	157	2.47
33	Nandi	1992	WB	Mixed	353	4.0	1424	6	4.21
34	Premarajan	1993	Pondicherry	Urban	225	4.7	1066	1	0.94
35	Sohi	1993	Chandigarh	Urban	2430	5.7	13,968	121	8.66
36	Shaji	1995	Kerala	Rural	1094	4.8	5284	27	5.11
37	Das	1996	WB	Rural	6500	5.8	37,286	114	3.06
38	Gourie-Devi	1996	Karnataka	Urban	550	5.5	3040	24	7.90

TABLE 7.1   *(Continued)*

Sl. No.	Chief investigator	Year of report	State/UT	Locality	No. of families	Average family size	No. of persons	No. of cases	Prevalence rate
39	Singh	1997	Haryana	Rural	–	–	30,000	126	4.20
40	Borah	1997	WB	Rural	–	–	8010	80	9.99
41	Mani	1997	Karnataka	Rural	–	–	64,936	416	6.41
42	Kokkat	1998	Haryana	Rural	–	–	8595	48	5.59
43	Nandi	2000a	WB	Rural	387	5.6	2183	15	6.87
44	Nandi	2000b	WB	Rural	506	6.9	3488	7	2.01
45	Radhakrishnan	2000	Kerala	Rural	43,681	5.5	23,8102	1175	4.94
46	Saha	2003	WB	Rural	3594	5.8	20,842	75	3.60
47	Gourie-Devi	2004	Karnataka	Mixed	–	–	10,2557	906	8.83

7.5.2 CODED MASTER SHEET

The codes used while classifying the Epilepsy studies of prevalence in India based on various attributes are presented in Table 7.2.

TABLE 7.2   Codes Used for Epilepsy Studies of Prevalence in India.

Study No.	Characteristics	Codes used	Sl. No	Characteristics	Codes used
1	**Domicile**		13	**Domicile & sex**	
	Rural			*Rural*	
		D1		Male	DSR1
	Urban	D2		Female	DSR2
	Semi-urban/Mixed			*Urban*	
		D3		Male	DSU1
2	**Region**			Female	DSU2
	Northern	R1	14	**Caste & sex (3)**	
	Eastern	R2		*Brahmins*	CSB1
				Male	
	Western	R3		Female	CSB2
	Southern	R4		*SC*	CSC1
				Male	

TABLE 7.2 *(Continued)*

Study No.	Characteristics	Codes used	Sl. No	Characteristics	Codes used
3	**Sex**			Female	CSC2
	Male	S1			
	Female	S2		*ST*	CST1
				Male	
4	**Age**			Female	CST2
	0–	A1	15	**Age & sex (1)**	
	1–	A2		0– Male	ASA1
	5–	A3		Female	ASA2
	10–	A4		5–Male	ASB1
	20–	A5		Female	ASB2
	30–	A6		10–Male	ASC1
	40–	A7		Female	ASC2
	50–	A8		20–Male	ASD1
	60–	A9		Female	ASD2
	70–	A10		30–Male	ASE1
5	**Marital status**			Female	ASE2
	Single	M1		40–Male	ASF1
	Married	M1		Female	ASF2
6	**Religion**			50–Male	ASG1
	Hindu	R1		Female	ASG2
	Muslim	R2		60–Male	ASH1
7	**Caste group**			Female	ASH2
	Brahmins	CG1	16	**Age & domicile (1)**	
	SC	CG2		0–9 age	
	ST	CG3		Rural	ADA1
	All others	CG4		Urban	ADA2
8	**Literacy level**			10–19 age	
	Illiterate/Primary	LL1		Rural	ADB1
	Secondary	LL2		Urban	ADB2
	University	LL3		20–29 age	
9	**Occupation**			Rural	ADC1
	Children	O1		Urban	ADC2
	Students	O2		30–39 age	

TABLE 7.2   *(Continued)*

Study No.	Characteristics	Codes used	Sl. No	Characteristics	Codes used
	House-wives	O3		Rural	ADD1
	All others	O4		Urban	ADD2
10	**Monthly income**			40–49 age	
	Low	MI1		Rural	ADE1
	Middle	MI2		Urban	ADE2
	High	MI3		49+ age	
11	**Family type**			Rural	ADF1
	Nuclear	FT1		Urban	ADF2
	Joint	FT2			
12	**Family size**				
	Up to 5	FS1			
	Above 5	FS2			

KEYWORDS

- **master sheet**
- **moderators**

REFERENCES

Cook, T. D.; Campbell, D. T. *Quasi-experimentation: Design and Analysis Issues for Field Settings*. Rand-McNally: Chicago, 1979.

Reddy, M. V. *Statistical Methods in Psychiatry Research*. Apple Academic Press Inc.: Canada, 2014.

CHAPTER 8

META-ANALYSIS PLOTS

CONTENTS

ABSTRACT

Forest plot is a graphical display of results from individual studies on a common scale. In this plot, the areas of the squares are proportional to the precision of the estimates and the lines joining the squares represent the confidence intervals. The funnel plot is a simple scatter plot of the treatment effects estimated from individual studies on the horizontal axis against the precision of the estimates on the vertical axis. The publication bias can be investigated by the use funnel plot. The Begg's funnel plot is a simple scatter plot of the standard error of the summary statistics on the horizontal axis against the summary statistics on the vertical axis. Galbraith plot is a graphical representation of the reciprocal of the standard error of each study estimate on the horizontal axis against the ratio of the summary statistic to its standard error (z-statistics). In Galbraith radial plot, the estimates from individual studies are plotted against the reciprocal of their standard errors. A L'Abbe plot is a scatter plot in which each point represents a study. With the vertical axis measuring the event rates in the treatment group and the horizontal axis the event rates in the control group. L'Abbe plot is useful to identify studies with differing results.

8.1   FOREST PLOT

It is a graphical display of results from individual studies on a common scale. In this plot, the areas of the squares are proportional to the precision of the estimates and the lines joining the squares represent the confidence intervals.

It allows a visual examination of the degree of heterogeneity between studies. It is used to display point estimates and corresponding confidence intervals for individual studies and the summary estimates. Individual studies may be displayed in different ordering in the Forest plot such as the year of study publication, estimated treatment effects, sample size, or severity of patients included. By doing this, the Forest plot may reveal systematic patterns about association of treatment effects with other study characteristics.

8.1.1 GRAPHICAL REPRESENTATION

The Forest plot prepared while studying a pattern and prevalence of epilepsy in India based on 47 selected studies to study the heterogeneity across studies is presented in Figure 8.1.

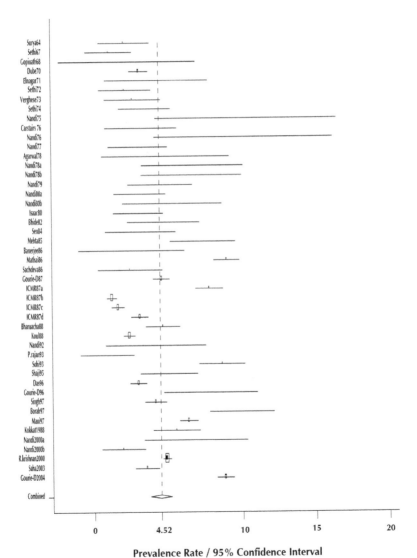

Prevalence Rate / 95% Confidence Interval

FIGURE 8.1   Forest plot depicting prevalence rates of epilepsy in India.

8.2   FUNNEL PLOT

The Funnel plot is a simple scatter plot of the treatment effects estimated from individual studies on the horizontal axis against the precision of the estimates on the vertical axis. Effect estimates from small studies will scatter more widely at the bottom of the graph, with the spread narrows among larger studies. In the absence of bias, the plot will resemble a symmetrical inverted funnel. Publication bias may lead to asymmetry in Funnel plot. The publication bias must be adjusted and the study results needed to be coded into a database and the criteria used in accepting or rejecting a study to be meta-analyzed need to be decided upon.

8.2.1   PUBLICATION BIAS

The publication bias can be investigated by the use Funnel plot. Perhaps the most common method to detect the existence of publication bias in a meta-analysis is the Funnel plot (Duval and Tweedie, 2000; Egger and Smith, 1997).

8.2.2   QUALITY OF DATA

Studies of lower quality also tend to show larger data points. In particular, studies with inadequate sampling method or less quality assessment scores may show inflated estimates. Smaller studies are, on average, conducted and analyzed with less methodological rigor than larger studies. The Funnel plot should be seen as a generic means of examining "small study effects" rather than a tool to diagnose specific types of bias.

8.2.3   ILLUSTRATION

The general Funnel plot prepared while studying a pattern and prevalence of epilepsy in India based on 47 selected studies to investigate publication bias is presented in Figure 8.2.

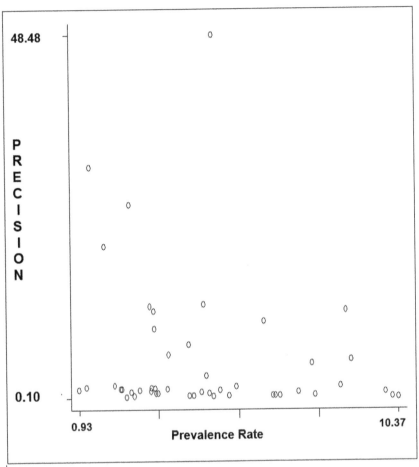

FIGURE 8.2 Funnel plot depicting prevalence studies of epilepsy in India.

8.2.4 BEGG'S FUNNEL PLOT

The Begg's Funnel plot is a simple scatter plot of the standard error of the summary statistics on the horizontal axis against the summary statistics on the vertical axis. This is compatible with a greater statistical power of the regression test. The horizontal line in this plot indicates the fixed effects summary estimate using IV method, while the sloping lines indicate the expected 95% confidence interval for a given standard error, assuming no heterogeneity between studies.

8.2.5 ILLUSTRATION

The Begg's Funnel plot prepared while studying a pattern and prevalence of epilepsy in India based on 47 selected studies to investigate publication bias is presented in Figure 8.3.

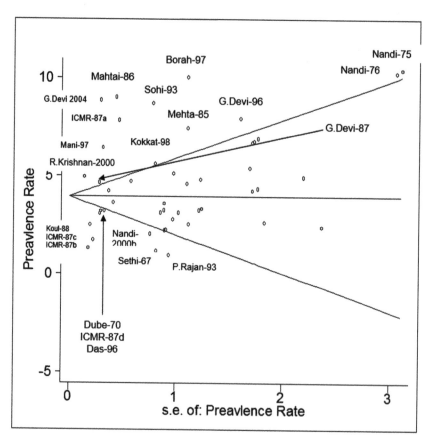

FIGURE 8.3   Begg's Funnel plot depicting prevalence studies of epilepsy in India.

8.3   GALBRAITH PLOT

It is a graphical representation of the reciprocal of the standard error of each study estimate on the horizontal axis against the ratio of the summary statistic to its standard error (z-statistics).

8.3.1   CLINICAL HETEROGENEITY AND STATISTICAL HETEROGENEITY

It was proposed to detect statistical heterogeneity. Clinical heterogeneity across the studies included in a meta-analysis is likely to lead to some degree of statistical heterogeneity in their results. The clinical heterogeneity is the differences between the studies in the patient selection, baseline disease severity, management of intermediate outcome, etc. Thus, the results of those studies were to some degree incompatible with one another. Such incompatibility in quantitative results is termed as statistical heterogeneity.

8.3.2   GALBRAIT RADIAL PLOT

In Galbraith radial plot, the estimates from individual studies are plotted against the reciprocal of their standard errors. This plot may help to judge visually which subset of the estimates are consistent with each other or with some theoretical value (Galbraith, 1988). The extent of heterogeneity across studies can be visually assessed according to whether individual studies scatter about a straight line through the origin. In this plot, less informative studies (with larger standard errors) cluster near the origin, whereas more informative studies (with smaller standard errors) are located further away from the origin.

8.3.3   ILLUSTRATION

The Galbraith plot prepared while studying a pattern and prevalence of epilepsy in India based on 47 selected studies as mentioned in the forest to detect statistical heterogeneity is presented in Figure 8.4.

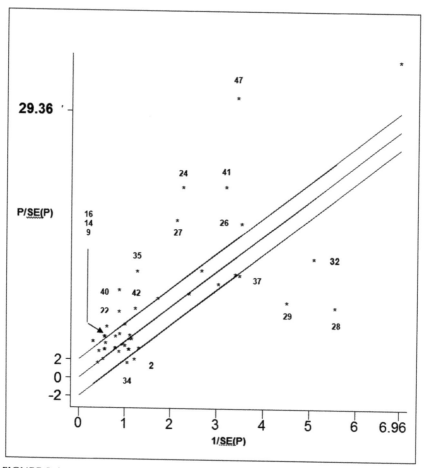

FIGURE 8.4   Galbraith plot depicting prevalence studies of epilepsy in India.

8.4   L'ABBE PLOT

A L'Abbe plot is a scatter plot in which each point represents a study, with the vertical axis measuring the event rates in the treatment group and the horizontal axis the event rates in the control group (L'Abbe et al., 1987).

8.4.1 MAIN PURPOSE

As with the Forest plot, L'Abbe plot is useful to identify studies with differing results. In addition, it enables identification of the study arms that are responsible for such difference. This may be important for determining the focus of heterogeneity investigations (Song et al., 2001). The investigation of heterogeneity in a meta-analysis may focus on individual study arms (either the intervention arms or the control arms or sometimes both arms) that show great variation across studies. Each point in the plot represents the results of study, with a larger point corresponding to a larger study. The solid diagonal line is the equal line on which the point of a study would lie when the event rate is same in the two groups. If a study point is below the equal line, it suggests that the rate of effect in treatment group is lower than that in the control group. The two dotted lines represent the overall weighted average effect by pooling the results of all studies (using the fixed or random effects model).

8.4.2 EXPLORING HETEROGENEITY

The L'Abbe plot is a useful tool, but it may also be used inappropriately, for example, in some meta-analysis, the L'Abbe plot has been used to identify outliers that were excluded one by one until a statistical test of heterogeneity was no longer statistically significant. In L'Abbe plot, random variation in the distance between a study point and the overall line is negatively related to the sample size, and the random variation in the distance is greatest when the event rates are 50% (Song et al., 2001). Therefore, to estimate the extent of extra heterogeneity (heterogeneity that cannot be explained by random variation), it is desirable to adjust the distances between a study point and the overall line by the study's sample size and event rate. It is a graphical representation of observed control group risk on the horizontal axis against observed intervention group risk on the vertical axis. It was proposed as a graphical means of exploring possible heterogeneity.

KEYWORDS

- **forest plot**
- **funnel plot**
- **Galbraith plot**
- **L'Abbe plot**

REFERENCES

Egger, M.; Smith, G. D. Meta-analysis: Potentials and Promise. *Br. Med. J.* **1997,** *315,* 1371–1374.

Duval, S.; Tweedie, R. Accounting for Publication Bias in Meta-analysis. *J. Am. Stat. Assoc.* **2000,** *95,* 89–98.

Galbraith, R. F. Graphical Display of Estimates Having Differing Standard Errors. *Technometrics* **1988,** *30,* 271–281.

L'Abbe, K. A.; Detsky, A. S.; O'Rourke, K. Meta-analysis in Clinical Research. *Ann. Intern. Med.* **1987,** *107,* 224–233.

Song, F.; Sheldon, T. A.; Sutton, A. J.; Abrams, K. R.; Jones, D. R. Methods for Exploring Heterogeneity in Meta-analysis. *Eval. Health Profess.* **2001,** *24,* 126–151.

CHAPTER 9

META-ANALYSIS OF TWO STUDIES

CONTENTS

ABSTRACT

While synthesizing the effect sizes of two separate studies, the meta-analyst seeks to compare the results to discover the degree of their actual similarity. Given two effect sizes that are not significantly different and therefore combinable on statistical grounds, you may want to determine the effect size of an effect across studies. The first step to take when combining the effect sizes of two studies is to calculate r for each and convert each r value into corresponding Z-scores. If the effect sizes of the two studies are statistically different, it means little sense to average their effect sizes. Although meta-analysts are usually more interested in effect sizes than p-values, they sometimes evaluate the overall level of significance as a way of increasing power. After we compare the results of two separate studies, it is an easy matter to combine the p-levels. In this way, we get an overall estimate of the probability that the two p-levels might have been obtained if the null hypothesis of no relation between X and Y were true.

9.1   COMBINING TWO STUDIES BY EFFECT SIZE

9.1.1   COMPARISON OF TWO STUDIES

In case of synthesizing, the effect sizes of two separate studies, the meta-analyst seeks to compare the results to discover the degree of their actual similarity (Rosenthal, 1984). The following steps are involved.

(1) Converting the quoted statistic from both studies, for example, "t" or chi square into "r"s.
(2) Give the calculated "r"s the same sign if both studies show effects in the same direction, but different signs if the results are in the opposite direction;
(3) Find for each "r" the associated "Fisher's z" value. Fisher's z refers to a set of log transformations of "r."
(4) Substitute in the following formula to find the Z score:

$$Z = \frac{z_1 - z_2}{\sqrt{\dfrac{1}{n_1 - 3} + \dfrac{1}{n_2 - 3}}}$$

(5) If the effect sizes produced by the two evaluated studies do not differ significantly, they are good candidates for combining. If a significant difference between effect sizes is found, we should investigate why the difference exists. You might look at the methods, materials, sample sizes, and procedures used in each study, as any or all of these may differ considerably between the studies and may be likely causes of the different effects.

9.1.2 COMBINING TWO STUDIES

Given two effect sizes that are not significantly different and therefore combinable on statistical grounds, you may want to determine the effect size of an effect across studies. The formula to be used again employs the Fisher's z-transformation:

$$\text{Mean } z \text{ or } z_m = \frac{z_1 + z_2}{2}$$

where z_1 and z_2 are Fisher's z scores.

The first step to take when combining the effect sizes of two studies is to calculate "r" for each and convert each "r" value into corresponding Z-scores. If the effect sizes of the two studies are statistically different, it means little sense to average their effect sizes.

If the results from the studies are in opposite direction, combining should never be considered.

9.2 COMBINING TWO STUDIES BY SIGNIFICANCE LEVEL

9.2.1 COMPARISON OF TWO STUDIES

Although meta-analysts are usually more interested in effect sizes than p-values, they sometimes evaluate the overall level of significance as a way of increasing power. It is again instructive to find out whether the individual values are homogeneous and therefore combinable.

9.2.2 TEST OF HOMOGENEITY

For each *p*-value, the meta-analyst then finds Z (i.e., not the Fisher *z*, but the standard normal deviate Z) using the table of Z. Both *p*-values should also be one-tailed, and we give the corresponding Z's the same sign if both studies showed effects in the same direction, but different signs if the results are in the opposite direction.

The difference between the two Zs' when divided by √2 yields a new Z.

This new Z corresponds to the *p*-value of the difference between the Z's if the null hypothesis were true (i.e., if the two Z's did not really differ).

Recapping,

$$Z = \frac{z_1 - z_2}{\sqrt{2}}$$

is distributed as Z, so we can enter this newly calculated Z in a table of standard normal deviates to find the *p*-value associated with a Z of the size obtained or larger.

9.2.3 COMBINING OF TWO STUDIES

After we compare the results of two separate studies, it is an easy matter to combine the *p*-levels. In this way, we get an overall estimate of the probability that the two *p*-levels might have been obtained, if the null hypothesis of no relation between X and Y were true. To perform these calculations, we modify the numerator of the formula for comparing *p*-values that we just described. We obtain accurate *p*-levels for each of our two studies and then find the Z corresponding to each of these *p*-levels. Also as before, both *p*'s must be given in one-tailed form, and the corresponding Z's will have the same sign if both studies show effects in the same direction and will have different signs if the results are in the opposite direction.

The only change in the previous equation is to add the Z values instead of subtracting them

$$Z = \frac{z_1 + z_2}{\sqrt{2}}.$$

This new Z corresponds to the p-value of the two studies combined if the null hypothesis of no relation were true.

KEYWORDS

- **two studies**
- **p-levels**
- **overall estimate**

REFERENCE

Rosenthal, R. *Meta-analytic Procedures for Social Research.* Sage: Beverly Hills, CA, 1984.

CHAPTER 10

METHODS FOR POOLING ESTIMATES: FIXED EFFECTS MODEL

CONTENTS

ABSTRACT

There are at least three sources of variation to consider before combining summary statistics across studies. They are inter-study variation, sampling error among studies, and study-level characteristics. Once the data have been assembled, simple inspection of the Forest plot is informative. Heterogeneity between study results should not be seen as purely a problem for systematic reviews, since it also provides an opportunity for examining why treatment effects differ in different circumstances. The sample size method used for meta-analysis employs a weighted average of the results in which the larger study generally has more influence than the smaller ones. Selection of a meta-analysis method for a particular analysis depends on the type of primary studies, choice of summary statistics, observed heterogeneity, the known limitations of the computational methods, and fixed effects versus random effects model. Fixed effects model is centered on making inferences for every population that have been sampled, then the outcomes are considered fixed and the only source of uncertainty is that resulting from the sampling of people into the studies. Pooling of study results under sample size method is mainly done under the assumption that, k samples are from a normal population. The inverse-variance method is used to pool binary, continuous, and correlation data. This approach has wide applicability since it can be used to combine any estimate that has standard error available. Mantel--Haenszel methods have been shown to be more robust when data are sparse, and may therefore be preferable to the inverse-variance method. In combining odds ratio, an alternative to the Mantel--Haenszel method is Peto.

10.1 SOURCE OF VARIATION

There are at least three sources of variation to consider before combining summary statistics across studies:

(1) Inter-study variation.
(2) Sampling error among studies.
(3) Study-level characteristics may differ among the studies.

There are a variety of statistical techniques available for accounting these different variations, which can be broadly classified into two models.

The difference between these models is the way the variability of the results between the studies is treated. The thoughtful consideration of heterogeneity between study results is an important aspect of systematic reviews. This should start when writing the review protocol, by defining potential sources of heterogeneity and planning appropriate subgroup analysis.

10.2 HETEROGENEITY

Once the data have been assembled, simple inspection of the Forest plot is informative. Statistical tests of homogeneity assess whether the individual study results are likely to reflect a single underlying effects, as opposed to a distinction of effects. If the test fails to detect heterogeneity among results, then it is assumed that the differences observed between individual studies are consequences of sampling variation and simply due to chance. The overview of the multicenter study suggests that a precondition for a valid meta-analysis is to test for the homogeneity of results. There are several statistical methods that allow the researcher to assess the magnitude of the problem. In fact, such tests are very informative, creating the possibility of identifying groups of studies with homogenous outcomes (Fisher et al., 1993).

10.2.1 TEST OF SIGNIFICANCE

A statistical test for the homogeneity of study means is equivalent to testing,

$H_0: \theta = \theta_1 = \theta_2 = \dots = \theta_k$ against H_1: At least one θ_i is different.
Under H_0, for large sample sizes,

$$Q_w = \sum_1^k w_i \left(\theta_i - \theta_{MLE} \right)^2$$

Follows Chi-square with $(k - 1)$ degrees of freedom

where $\theta_{MLE} = \sum w_i \, \theta_i / \sum w_i$ and $w_i = 1/s_i^2$

If H_0 is rejected, the meta-analysis may conclude that the study means arose from two or more distinct populations. If H_0 cannot be rejected, the investigator would conclude that the studies share a common mean θ, and estimate θ using θ_{MLE}. This test of homogeneity has low power to detect heterogeneity, and therefore it is advisable to take a higher significance level than usual, that is for example 10%.

10.2.2   REMARKS

Heterogeneity between study results should not be seen as purely a problem for systematic reviews, since it also provides an opportunity for examining why treatment effects differ in different circumstances. Careful investigations of heterogeneity in meta-analysis should increase the scientific and clinical relevance of their results. Kulinskaya et al. (2004) have devised a welch-type test for homogeneity of contrast under heteroscedasticity with application to meta-analysis.

10.3   PRINCIPLES OF POOLING ESTIMATES

10.3.1   IMPORTANCE OF SENSITIVITY ANALYSIS

There are different statistical methods for combining the data in a meta-analysis, but there is no single correct method. A thorough sensitivity analysis should always be performed to assess the robustness of combined estimates to different assumptions, methods, and inclusion criteria and to investigate the possible influence of bias.

10.3.2   SAMPLE SIZE

The principle that simply pooling the data from different studies and treating them as one large study would fail to preserve the randomization and introduce bias and confounding. The results from small studies are more subjective to the play of chance and should therefore be given less weight. Hence, the methods used for meta-analysis employ a weighted average of the results in which the larger studies generally have more influence than the smaller ones.

10.3.3   CHOICE OF METHOD

Selection of a meta-analysis method for a particular analysis depends on the type of primary studies, choice of summary statistics, observed hetero-geneity, the known limitations of the computational methods, and fixed effects versus random effects model. It would be a great value if, for each study entering into a meta-analysis, a single summary statistic could be computed that incorporate the information from all the effect sizes rele-vant to the hypothesis being tested (Rosenthal, 1984).

10.4   FIXED EFFECTS MODEL

If interest is centered on making inferences for every populations that have been sampled, then the outcomes are considered fixed and the only source of uncertainty is that resulting from the sampling of people into the studies. This type of variation may be characterized as within-study variation that is function of the number of people in the primary study and the vari-ability in the people's responses within the primary studies. In this case, a fixed-effects model would be used for statistical inference. Inferences are similar to those made when performing an ANOVA when there is no inter-study variation in the mean outcome. The intuition underpinning the fixed effects model is that other levels of treatments are sufficient like those in the sample of primary studies that inferences would be same. The popu-lation to which generalization are to be made consists of a set of studies having identical characteristics and study effects (Normand, 1999). Thus the fixed effects model assumes the variability between studies as due to random variation and individual studies are simply weighted by their precision. Therefore, if all the studies were infinitely large, they would give identical results.

10.5   SAMPLE SIZE METHOD

Pooling of study results under fixed effects method is mainly done under the assumption that, k samples are from a normal population (Williamson et al., 2002). Suppose n_j from study j ($j = 1,2,...,k$) are diagnosed for particular disorder for individual i ($i = 1,2,..., n_j$, in study j). We assume that, prevalence rates p_j are from a normal population with unknown mean

θ_j and variance σ_j^2. To pool the results under a fixed effects model, we assume $\theta_j = \theta$ and $\sigma_j^2 = \sigma^2$ for all j. To estimate θ and σ^2 standard methods (Williamson et al. 2002) for the pooling of k samples from the same normal population are as follows:

$$\hat{\theta} = \frac{\sum_{j=1}^{k} n_j p_j}{\sum_{j=1}^{k} n_j} \quad \text{with its standard error,} \quad \hat{\sigma}^2 = \frac{\sum_{j=1}^{k} (n_j - 1) s_j^2}{\sum_{j=1}^{k} (n_j - 1)}$$

10.6 INVERSE-VARIANCE METHOD

The inverse-variance method (IV method) is used to pool binary, continuous, and correlation data. This approach has wide applicability, since it can be used to combine any estimate that has standard error available. The effect sizes are combined to give a pooled estimate (denoted by θ) by calculating weighted average of the treatment effects from the individual studies as follows.

$$\theta_{IV} = \frac{\sum w_i \theta_i}{\sum w_i}$$

where the weights w_i are calculated as

$$w_i = \frac{1}{SE(\theta_i)^2}$$

That is, the weight for the ith study is equal to its precision of the estimate. The standard error of is θ_{IV} given by

$$SE(\theta_{IV}) = \frac{1}{\sqrt{\sum w_i}}$$

The heterogeneity statistic is given by

$$Q_w = \sum w_i (\theta_i - \theta_{IV})^2$$

The Q_w follows chi-square distribution with $(k-1)$ degrees of freedom, where k is the number of studies included in the meta-analysis.

10.7 MANTEL–HAENSZEL METHOD

When data are sparse, both in terms of event rates being low and trials being small, the estimates of the standard errors of the treatment effects that are used in the inverse variance methods may be poor. Mantel–Haenszel methods use an alternative weighting scheme, and have been shown to be more robust when data are sparse, and may therefore be preferable to the inverse-variance method. In other situations, they give similar estimates to the inverse variance method. They are available only for binary outcomes.

For each study, the effect size from each trial θ_i is given weight w_i in the analysis. The overall estimate of the pooled effect θ_{MH} is given by

$$\theta_{MH} = \frac{\sum w_i \theta_i}{\sum w_i}$$

Unlike with inverse variance methods, relative effect measures are combined in their natural scale, although their standard errors are still computed on log scale.

10.7.1 COMBINING ODDS RATIO

ith study's OR is given weight based on the following date structure for each study.

Study	Event	No Event	Group Size
Intervention	a	b	n_1
Control	c	d	n_2
Total			N

$$w = \frac{bc}{N}$$

and the logarithm of OR$_{MH}$ has standard error given by

$$SE(Ln(OR_{MH})) = \sqrt{\frac{1}{2}\left(\frac{E}{R^2} + \frac{F+G}{R \times S} + \frac{H}{S^2}\right)}$$

where

$$R = \sum \frac{a/d}{N}$$

$$S = \sum \frac{bc}{N}$$

$$E = \sum \frac{(a+d)a/d}{N^2}$$

$$F = \sum \frac{(a+d)b/c}{N^2}$$

$$G = \sum \frac{(b+c)a/d}{N^2}$$

$$E = \sum \frac{(b+c)b/c}{N^2}$$

10.7.2 COMBINING RISK RATIO

For combining risk ratio, each study's RR is given weight:

$$w = \frac{c\,n_1}{N}$$

and the logarithm of RR$_{MH}$ has standard error given by

$$SE(Ln(RR_{MH})) = \sqrt{\frac{P}{R \times S}}$$

where

$$P = \sum \frac{\left(n_1 / n_2 \left((a+c) - acN\right)\right)}{N^2}$$

$$R = \sum \frac{an_2}{N}$$

$$S = \sum \frac{cn_1}{N}$$

10.7.3 COMBINING RISK DIFFERENCE

For risk difference, each study's RD has the weight:

$$w_i = \frac{n_1 n_2}{N}$$

and RD_{MH} has standard error given by

$$SE(RD_{MH}) = \sqrt{\frac{J}{K^2}}$$

where

$$J = \sum \left(\frac{abn_2^3 + c_i d_i n_1^3}{n_1 n_2 N^2} \right)$$

$$K = \sum \frac{n_1 n_2}{N}$$

However, the test of homogeneity is based upon the Inverse-variance weights and not the Mantel–Haenszel weights. The heterogeneity statistic is given by

$$Q_w = \Sigma w_i \left(\theta_i - \theta_{MH} \right)^2$$

where θ_i is the log odds ratio, log relative risk, or risk difference.

10.8   PETO METHOD

10.8.1   COMBINING ODDS RATIO

An alternative to the Mantel–Haenszel method is a method due to Peto. The overall odds ratio is given by

$$OR_{Peto} = \exp\left(\frac{\sum w_i \ln(OR_i)}{\sum w_i} \right)$$

where the odds ratio OR_i is calculated using the approximate Peto method described in individual trial, and the weight w_i is equal to the hypergeometric variance of the event count in the intervention group, v_i.

The logarithm of the odds ratio has standard error:

$$SE(\ln(OR_{Peto})) = \frac{1}{\sqrt{\sum v_i}}$$

The heterogeneity statistic is given by

$$Q = \sum v_i \left(\ln\left(OR_i\right) - \ln\left(OR_{Peto}\right) \right)^2$$

The approximation upon which Peto's method rallies has shown to fail, (1) when treatment effects are very large and (2) sizes of the arms of the trials are seriously unbalanced.

Severe imbalances, with, example four or more times as many participants in one group than the other, would rarely occur in randomized trials. In other circumstances, including when event rates are very low, the method performs well. Corrections for zero cell counts are not necessary for this method.

KEYWORDS

- **fixed effects model**
- **sample size method**
- **inverse-variance method**
- **Peto**

REFERENCES

Fisher, L. D.;, Belle, G. V. *Biostatistics : A Methodology for the Health Sciences*. John Wiley and Sons Inc. : New York, 1993.

Kulinskaya, E.,; Dollinger, M. B.,; Knight, E.,; Gao H. A Welch-type Test for Homogeneity of Contrasts under Heteroscedasticity with Application to Meta-analysis. *Stat. Med.* **2004,** *23*, 3655–3670.

Normand, S. T. Meta-analysis: Formulating, Evaluating, Combining, and Reporting. *Stat. Med.* **1999,** *18*, 321–359.

Rosenthal, R. *Meta-analytic Procedures for Social Research*. Sage: Beverly Hills, CA, 1984.

Williamson, P. R.; Lancaster, G. A.; Craig, J. V.; Smyth, R. L. Meta-analysis of Method Comparison Studies. *Stat. Med.* **2002,** *21*, 2013–2025.

CHAPTER 11

METHOD FOR POOLING ESTIMATES: RANDOM EFFECTS MODEL

CONTENTS

ABSTRACT

The fixed-effect model is based on the assumption that there is one true effect size which is shared by all the included studies. The random-effects model is based on the assumption that the true effect could vary from study to study. If inferences were to be generalized to a population in which the studies are permitted to have different effects and different characteristics, then a random-effects model would be appropriate. The random-effects model leads to relatively more weights being given to smaller studies and to wider confidence intervals than the fixed effects model. The DerSimonian and Laird method of meta-analysis is based on the random-effects model with the assumption of common effect is relaxed and the effect sizes are assumed to have a normal distribution with mean θ and variance τ^2. The maximum likelihood method is considered when variance of the estimator is assumed as known. Restricted maximum likelihood method is an approach to estimation that maximizes the likelihood over a restricted parameter space. Applying Bayesian methods to perform meta-analysis provides a more informative summary of the data than non-Bayesian approaches. The major advantage is the ability to incorporate the uncertainty from our estimates of the true effects in individual studies. The practice of starting with the fixed-effect model and then moving to a random-effect model, if Q is statistically significant should be discouraged. If the logic of the analysis says that we are trying to estimate a range of effects, then the selection of a computational model should be based on the nature of the studies and our goal.

11.1   RANDOM-EFFECTS MODEL

If inferences were to be generalized to a population in which the studies are permitted to have different effects and different characteristics, then a random-effects model would be appropriate. The intuition underpinning random-effects models is that because there are many different approaches to conducting a study by perturbing the design in a small way, and then there are many different potential treatment effects that could arise. This situation corresponds to an ANOVA model in which there is inter-study variation in the mean outcome in addition to the within-study variation. Thus, the population in a random-effects model is the one in which there are infinitely many possible populations (Normand, 1999). Thus, the random-effects model assumes different underlying effects for each study and takes this into consideration as an additional source of variation.

The random-effects model leads to relatively more weights being given to smaller studies and to wider confidence intervals than the fixed-effects model. The between study variation τ^2 places an important role and must also be estimated. When τ^2 is zero the random-effects model corresponds to the fixed-effects model. With binary responses, random and fixed-effect assumptions may lead to very different conclusions, so that one is no longer an alternative to the other (Lee, 2002).

11.2   METHODS FOR POOLING ESTIMATES

11.2.1   DERSIMONIAN AND LAIRD METHOD

The DerSimonian and Laird method (DL method) of meta-analysis is based on the random-effects model. Under the random-effects model, the assumption of common effect is relaxed, and the effect size θ_i are assumed to have a normal distribution with mean θ and variance τ^2. The usual DL estimate for τ^2 is given by

$$\tau^2 = \frac{Q_w - (k-1)}{\sum w_i - \left(\sum w_i^2 / \sum w_i\right)}$$

where Q_w is the heterogeneity statistic, and the weights w_i are calculated as in the inverse variance method, and k is the number of studies. The τ^2 is set to zero if $Q_i < (k-1)$. In this approach, the weights for each study-effect size are as given below.

$$w_i' = \frac{1}{\mathrm{SE}(\theta_i)^2 + \tau^2}$$

The pooled estimate is given by

$$\theta_{DL} = \frac{\sum w_i' \theta_i}{\sum w_i'} \quad \text{and} \quad \mathrm{SE}(\theta_{DL}) = \frac{1}{\sqrt{\sum w_i'}}$$

The heterogeneity statistic and its test of significance are as given in the IV method. The weights in this method will be smaller and more similar to each other than the weights in IV method.

When combining results from separate investigations in a meta-analysis, random-effects methods enable the modeling of difference between studies by incorporating a heterogeneity parameter τ^2 that accounts explicitly for cross-study variation (Biggerstaff and Tweedie, 1997; Johnson and Thompson, 1995).

11.2.2   MAXIMUM LIKELIHOOD METHOD

It is shown that the commonly used DL method does not adequately reflect the error associated with parameter estimation (Brockwell and Gordon, 2001). The maximum likelihood estimation (MLE) method is considered when variance of the estimator is assumed as known. Then the log likelihood of the estimator will yield the MLE (Normand, 1999). The MLE being an iterative scheme for estimating τ^2, it has to be estimated by solving the iterative equation (Hardy and Thompson, 1996). According to Thompson and Sharp (1999), MLE for τ^2 is given by

$$\tau^2 = \frac{\sum w_i^{*2}\left(\left(\theta_i - \hat{\theta}\right)^2 - v_i\right)}{\sum w_i^{*2}}$$

where $w_i^* = \dfrac{1}{s_i^2 + \tau^2}$

Starting with $\tau^2 = 0$, gives initial value for $\hat{\theta}$ which is equal to θ_{IV}. This in turn will yield a new value for τ^2 (subject to the constraint that negative values are set to zero). This provides for modified weights w_i^*, leading to new estimate of τ^2. The procedure continues until convergence taken place.

11.2.3   RESTRICTED MAXIMUM LIKELIHOOD METHOD

Restricted maximum likelihood estimation (REML) is an approach to estimation that maximizes the likelihood over a restricted parameter space.

While applicable to more general models, it has most often been applied to the estimation of variance components in a general linear model with a multivariate normal distribution. It is an alternative to maximum likelihood (ML) estimation, which leads to unbiased estimators.

Essentially, the procedure adjusts for the fact that fixed effects are unknown when estimating components of variance. In balanced analysis of variance settings, this takes the form of adjustment in degrees of freedom. In these settings, the REML estimators of the variances are the familiar unbiased least square estimators (Cook, 1998).

Patterson and Thompson (1971) introduced REML as a method of estimating variance components in the context of unbiased incomplete block designs. REML is similar to the ML method, but it first separates the likelihood into two parts: one that contains the fixed effects and the other that does not.

The REML is often preferred to MLE because it takes account of the loss of degree of freedom in estimating mean and produces unbiased estimating equations for the variance parameters (Smyth and Verbyla, 1996). The REML method is for estimating variance components in a general linear model using the marginal distribution of y where log-likelihood is to be maximized (Normand, 1999).

The use of REML estimates overcomes the tendency of MLE method to underestimate variances. In this context, the REML estimate for τ^2 is given by (Thompson and Sharp, 1999) allowing a correction factor as per number of parameters to be estimated.

$$\tau^2 = \frac{\sum w_i^{*2}\left(\frac{k}{k-1}\right)\left(\left(\theta_i - \hat{\theta}\right) - v_i\right)}{\sum w_i^{*2}}$$

where θ_i and v_i are the end point estimate and its variance of ith study.

The REML method deals with linear combinations of the observed values whose expectations are zero. These "error contrasts" are free of any fixed effects in the model. In contrast to maximize likelihood estimates, REML estimates of variances and co-variances are known to be unbiased (Bob-Baker, 2002).

11.2.4   EMPIRICAL BAYESIAN METHOD

An estimator for θ_i can be calculated by substituting the REML estimates for the hyper-parameters. This type of approximation to the posterior distribution is known as empirical Bayes (Normand, 1999) has concentrated on Empirical Bayes estimate for obtaining τ^2 by replacing w_i^{*2} with w_i^* in REML estimate of τ^2 and provided the following formula (Thompson and Sharp, 1999).

$$\frac{\sum w_i^* \left(\frac{k}{k-1} \right) \left(\left(\theta_i - \hat{\theta} \right) - v_i \right)}{\sum w_i^*}$$

Applying Bayesian methods to perform meta-analysis provides a more informative summary of the data than non-Bayesian approaches. Two major advantages are the ability to incorporate the uncertainty from our estimates of the true effects in individual studies. Standard methods assume either no between-study heterogeneity (fixed-effects models) or use the most likely value of the between-study variance τ^2 (random-effects models). Both approaches ignore large values than the data might support for τ^2, which might substantially change the weighting of the different studies. Although meta-analysis is often focused on estimating the common study mean, evaluation of heterogeneity is a crucial part of a meta-analysis (Antman et al., 1996). Obtaining posterior estimates of study effects can aid in determining whether studies really are heterogeneous or whether perceived heterogeneity is an artifact of small sample sizes. Further exploration through meta-regression or individual patient regression may help uncover important treatment-effect modifiers and causes of between-study heterogeneity (Duangjinda et al., 2001; Houwelingen et al., 2002; Leonard and Duffy, 2002; Schmid, 2001; Smith et al., 1995). Nam et al. (2003) have proposed and evaluated three Bayesian multivariate meta-analysis models.

11.3   DIFFERENCE BETWEEN FIXED AND RANDOM-EFFECTS MODEL

The fixed-effect model is based on the assumption that there is one true-effect size which is shared by all the included studies. The random-effects model is based on the assumption that the true effect could vary from study to study. Consequently, in fixed-effects model, a large study would give the lion's share of the weight and the small study could be largely ignored. In random-effects model, larger studies are less likely to dominate the analysis and small studies are less likely to be trivialized.

11.3.1 CHOICE OF THE MODEL

The practice of starting with the fixed-effect model and then moving to a random-effect model if Q is statistically significant should be discouraged. If the logic of the analysis says that we are trying to estimate a range of effects, then the selection of a computational model should be based on the nature of the studies and our goal. When the researcher is accumulating data from a series of studies that had been performed by other people, it would be very unlikely that all the studies were functionally equivalent.

If the study effect sizes are seen as having been sampled from a distribution of effect sizes, then the random-effects model, which reflects this idea, is the logical one to use. If the between-studies variance is statistically significant, then the fixed-effect model is inappropriate. However, even if the between-studies variance does not meet the criterion for statistical significance, we should still take account of this variance when assigning weights.

On the other hand, if one has elected to use the fixed-effect model a priori but the test of homogeneity is statistically significant, then it would be important to revisit the assumptions that led to the selection of a fixed-effect model.

KEYWORDS

- **DerSimonian and Laird method**
- **maximum likelihood method**
- **restricted maximum likelihood method**
- **Bayesian method**

REFERENCES

Antman, E. M.; Seelig, M. S.; Fleishman, K.; Lau, J.; Berkey, C. S.; McLntosh, M. W. Magnesium is Acute Myocardial Infarction: Scientific, Statistical and Economical Reasons for its Use. *Cardiovasc. Drugs Ther.* **1996**, *10*, 297–301.

Biggerstaff, B. J.; Tweedie, R. L. Incorporating Variability in Estimates of Heterogeneity in the Random Effects Model in Meta-analysis. *Stat. Med.* **1997**, *16*, 753–768.

Bob-Baker, R. J. *Basic Principles of Statistical Analysis (Chapter 4 of on Line Book).* 2002, http://duke.usask.ca/~rbaker/stats.html.

Brockwell, S. E.; Gordon, I. R. A Comparison of Statistical Methods for Meta-analysis. *Stat. Med.* **2001**, *20*, 825–840.

Cook, N. R. Restricted Maximum Likelihood. In *Encyclopedia of Biostatistics*; Armitage, P., Colton, T., Eds.; John Wiley and Sons: New York, 1998; Vol 5, pp 3827–3630.

Duangjinda, M.; Misztal. I.; Bertrand, J. K.; Tsuruta, S. The Empirical Bias of Estimates by Restricted Maximum Likelihood, Bayesian Method, and Method R under Selection for Additive, Maternal, and Dominance Models. *J. Anim. Sci.* **2001**, *79*, 2991–2996.

Hardy, R. J.; Thompson, S. G. A Likelihood Approach to Meta-analysis with Random Effects. *Stat. Med.* **1996**, *15*, 619–629.

Houwelingen, H. C. V.; Arends, L. R.; Stijnen, T. Advanced Methods in Meta-analysis: Multivariate Approach and Meta-regression. *Stat. Med.* **2002**, *21*, 589–624.

Lee, Y. Fixed-effect versus Random-effect Models for Evaluating Therapeutic Preferences. *Stat. Med.* **2002**, *21*, 2325–2330.

Leonard, T.; Duffy, J. C. A Bayesian Fixed Effects Analysis of the Mantel–Haenszel Model Applied to Meta-analysis. *Stat. Med.* **2002**, *21*, 2295–2312.

Johnson, D. L.; Thompson, R. Restricted Maximum Likelihood Estimation of Variance Components for Univariate Animal Models Using Sparse Matrix Techniques and Average Information. *J. Dairy Sci.* **1995**, *78*, 499–456.

Nam, I. S.; Mengersen, K.; Garthwaite, P. Multivariate Meta-analysis. *Stat. Med.* **2003**, *22*, 2309–2333.

Normand, S. T. Meta-analysis: Formulating, Evaluating, Combining, and Reporting. *Stat. Med.* **1999**, *18*, 321–359.

Patterson, H. D.; Thompson, R. Recovery of Inter-block Information When Block Sizes are Unequal. *Biometrika* **1971**, *58*, 545–554.

Schmid, C. H. Using Bayesian Inference to Perform Meta-analysis. *Eval. Health Profess.* **2001**, *24*, 165–189.

Smith, T. C.; Spiegelhalter, D. J.; Thomas, A. Bayesian Approaches to Random-effects Meta-analysis: A Comparative Study. *Stat. Med.* **1995**, *14*, 2685–2699.

Smyth, G. K.; Verbyla, A. P. A Conditional Approach to Residual Maximum Likelihood Estimation in Generalized Linear Models. *J. R. Stat. Soc. B* **1996**, *58*, 565–572.

Thompson, S. G.; Sharp, S. J. Explaining Heterogeneity in Meta-analysis: A Comparison of Methods. *Stat. Med.* **1999**, *18*, 2693–2708.

CHAPTER 12

INDIVIDUAL PATIENT DATA META-ANALYSIS

CONTENTS

ABSTRACT

The predominant difference between an individual patient data meta-analysis (IPD) and meta-analysis based on aggregate data is that the combined study results from a central re-analysis of the raw data from each study. The IPD meta-analyses involve the collaboration of the investigators. Analyst can use IPD to examine and update trial data and to carry out time-to-event analysis. The summary statistics can be calculated for specific groups of patients and thus subgroup analysis can also be produced. The problem of ecological fallacies may be avoided in meta-analysis using IPD. Ecological fallacy refers to a situation where an average estimate for a group is not valid to the individual patients. Multivariate regression analysis using IPD may provide results that are directly relevant to individual patients. However, meta-analysis using individual patient data is much more expensive to conduct.

12.1   FEATURES OF INDIVIDUAL PATIENT DATA META-ANALYSIS

The predominant difference between an individual patient data meta-analysis (IPD) and meta-analysis based on aggregate data is that the combined study results from a central reanalysis of the raw data from each study. The necessary data items are sought and, after central processing, any inconsistencies or problems are discussed and hopefully resolved by communication with the responsible investigators. The finalized data for each study summary statistics, which are combined to give an overall estimate of the effect of treatment. In this way, participants are only directly compared with others in the same study and the entire dataset is not pooled as though it came from a single homogeneous study.

The IPD meta-analysis is based on the basic approaches and methods of meta-analysis. The IPD meta-analysis calculates the common summary tables and statistics for each study and the same method of analysis is used for all the studies.

12.2   ADVANTAGES OF IPD META-ANALYSIS

12.2.1   COMMUNICATION WITH INVESTIGATOR

The IPD meta-analyses involve the collaboration of the investigators. These include more complete identification and understanding of the studies, better compliance with providing missing data, more balanced interpretation of the results of the review, wider endorsement and dissemination of these results, a broader consensus on the implications for future practice and research, and possible collaboration in such research. The quality of the randomization in meta-analysis is ensured through detailed data checking, and iterative correction of errors by communication with the investigators. The basic data used in meta-analysis is updated by follow-up information through patient record systems.

12.2.2   TIME TO EVENT DATA

Analyst can use IPD to examine and update trial data and to carry out time-to-event analysis.

12.2.3   SUBGROUP META-ANALYSIS

The summary statistics can be calculated for specific groups of patients and thus subgroup analysis can also be produced.

12.2.4   AVOIDING ECOLOGICAL FALLACIES

The problem of ecological fallacies may be avoided in meta-analysis using IPD. Ecological fallacy refers to a situation where an average estimate for a group is not valid to the individual patients.

12.2.5   MULTIVARIATE ANALYSIS

Multivariate regression analysis using IPD may provide results that are directly relevant to individual patients (Lau et al., 1998).

12.2.6   WIDE APPLICATIONS

The IPD meta-analysis has been used in large-scale collaborative overviews in which data from all randomized trials in a particular disease area are brought together and also in more restricted reviews in which data from a relatively small number of trials assessing a specific healthcare intervention are collected and combined.

12.2.7   INTENTION-TO-TREAT PRINCIPLE

As with any systematic review the fundamental principle for one which uses IPD is that as much as possible of the relevant, valid evidence is included. This means that the process of identification must be as thorough as possible and that the attempts to collect data must be equally thorough. The ultimate aim should be that all randomized participants, and no non-randomized participants, from all relevant studies are included and that they are analyzed using the intention-to-treat principle. In this way, systematic biases and chance effects will be minimized.

12.2.8   DATA COLLECTION

The data collection should be kept simple and straightforward, with the minimum amount of data being collected for the required analyses. It should be as easy as possible for the investigators to supply their data, since this should increase the likelihood that data will be received for all relevant studies.

Investigators should know that data supplied for the review will be held in confidence and will not be used for any other purpose without their permission, and that the reports of the review will be published in the names of the collaborating investigators rather than the central cocoordinators.

12.3   LIMITATIONS

However, meta-analysis using IPD is much more expensive to conduct. If important covariates have not been collected in primary studies, investigation of heterogeneity will also be limited in meta-analysis using IPD.

KEYWORDS

- **IPD**
- **time-to-event analysis**
- **subgroup analysis**
- **regression analysis**

REFERENCE

Lau, J.; Loannidis, J. P.; Schmid, C. H. Summing Up Evidence: One Answer is Not Always Enough. *Lancet* **1998,** *351,* 123–127.

CHAPTER 13

META-ANALYSIS OF OBSERVATIONAL STUDIES

CONTENTS

ABSTRACT

Meta-analysis of observational studies is also common in psychiatry research. Observational studies yield estimates of association that may deviate from true underlying relationship beyond the play of chance. This may be due to the effects of confounding factors, bias or both. Consideration of possible sources of heterogeneity between observational studies results will provide more insights than the mechanistic calculation of an overall measure of effect, which may often be biased. Meta-analysis of evaluation of prognostic variables has a higher risk of missing studies than for randomized trials. Most of the prognostic studies are found to be methodologically poor. It is more difficult to identify all prognostic studies by literature search than for randomized trials. Many studies seek parsimonious prediction models by retaining only the most important prognostic factors. If the prognostic variable is continuous, the risk of an event would usually be expected to increase or decrease systematically as the level increases. The method of analysis for pooling values across several studies will depend on whether the prognostic variable is binary, categorical, or continuous. In case of time to event data, the data is analyzed using survival analysis method-most often the log-rank test for simple comparisons on Cox regression for analysis of multiple predictor variables. Some meta-analysis studies consider the set of research studies where the aim was to investigate many factors simultaneously to identify important risk factors. Statistical power is rarely discussed in studies of diagnostic accuracy as they do not compare two groups, and they do not formally test hypotheses. The choice of a statistical method for pooling results depends on the source of heterogeneity, especially variation in diagnostic thresholds. There is also one important extra source of variation to consider in meta-analysis of diagnostic accuracy: variation introduced by changes in diagnostic threshold. Sensitivities and specificities, and positive and negative likelihood ratios, can be combined into the same single summary of diagnostic performance, known as the diagnostic odds ratio. If there is any evidence that the diagnostic threshold varies between the studies, the best summary of the results of the studies will be an ROC curve rather than a single point. The simplest method of combining studies of diagnostic accuracy is to compute weighted averages of the sensitivities, specificities, or likelihood ratios. Likelihood ratios are ratios of probabilities and in a meta-analysis can be treated as risk ratio. A weighted average of the likelihood ratios can be computed using the standard Mantel–Haenszel or inverse variance methods of meta-analysis of risk ratios.

13.1   OBSERVATIONAL META-ANALYSIS STUDIES

The previous chapters dealt with meta-analysis aspect of randomized trials. Meta-analysis of observational studies is also common in psychiatry research. Most of them are based on cohort and case-control studies (Egger et al., 2001).

13.1.1   NEED FOR OBSERVATIONAL STUDIES

Etiological hypothesis cannot generally be tested in randomized experiments. The evidence that is available from clinical trials will rarely answer many important questions. Such an adverse effect occurring later will not be identified; women, the elderly, and minority ethnic groups are often excluded from randomized trials. It is impossible to recruit sufficient patients into control trials. Meta-analysis may be attractive to reviewers in etiological epidemiology and observational effectiveness research.

13.1.2   CONFOUNDING AND BIAS

In observational studies, which yield estimates of association that may deviate from true underlying relationship beyond the play of chance. This may be due to the effects of confounding factors, bias, or both.

13.1.3   SPURIOUS FINDINGS

Epidemiological studies produce a large number of specific associations but some of them will be spurious. Bigger being better is fallacy in observational studies. The main problem is not lack of precision but that some studies produce findings that are seriously biased or confounded. Meta-analysis of observational studies may often unknowingly produce light confidence intervals around biased results.

13.1.4   *EXPLORING SOURCES OF HETEROGENEITY*

The statistical combination of studies should not be a prominent component of meta-analysis of observational studies. Through consideration of possible sources of heterogeneity between observational studies results will provide more insights than the mechanistic calculation of an overall measure of effect, which may often biased.

13.2   EVALUATION OF PROGNOSTIC VARIABLES

13.2.1   *FEATURES*

Meta-analysis of evaluation of prognostic variables has a higher risk of missing studies than for randomized trials. Prognostic variables should be evaluated in a representative sample of patients assembled at a common point in the course of the disease. Ideally, they should all have received the same medical treatment (Egger et al., 2001).

13.2.2   *IMPORTANCE OF IPD*

Most of the prognostic studies are found to be methodologically poor, in particular in relation to the analysis of continuous prognostic variables. Meta-analysis of prognostic studies using individual patient data (IPD) may overcome many of these difficulties. Access to IPD is therefore highly desirable to allow comparable analysis across studies.

13.2.3   *DETERMINATION OF PROGNOSTIC VARIABLES*

Prognostic studies take various forms. Some studies investigate the prognostic value of a particular variable, while others investigate many variables simultaneously. To evaluate which are prognostic are to develop a prognostic model for making prognosis for individual patients. It is not always easy to discern the aims of a particular study. Also, substudies are carried out to try to identify variables that predict response to treatment.

Prognostic studies include clinical studies of variables predictive of future events as well as epidemiological studies of etiological or risk factors. As multiple similar studies accumulate, it becomes increasingly important to identify and evaluate all of the relevant studies to develop a more reliable overall assessment.

13.2.4   IDENTIFICATION OF PUBLICATIONS

It is more difficult to identify all prognostic studies by literature search than for randomized trials. There is no optimal strategy for searching the literature for such studies.

13.2.5   INADEQUATE INFORMATION

Some authors fail to present the numerical summary of the prognostic strength of an available such as hazard ratio especially when the effect of variable was not statistically significant. Even when numerical results are given, they may vary in formats: for example survival proportions may be given for different time-points. Also, odds ratios or hazard ratios from grouped or ungrouped analysis are not comparable. Quantitative synthesis (meta-analysis) is not possible because the published papers do not all include adequate information.

13.2.6   ASSESSING METHODOLOGICAL QUALITY

There is little empirical evidence to support the importance of particular study features affecting the reliability of study findings, including the avoidance of bias. Several methodological considerations issues are similar to those relevant to studies of diagnosis.

13.2.7   OTHER PROGNOSTIC VARIABLES

It is important to adjust for other prognostic variables to get a valid picture of the relative prognosis for different values of the primary prognostic variable. It is necessary because patients with different values/covariates of primary interest are likely to differ with respect to other prognostic variables.

13.2.8 PREDICTION MODEL

Many studies seek parsimonious prediction models by retaining only the most important prognostic factors, most commonly by using multiple regression analysis with step wise variable selection.

13.2.9 HANDLING CONTINUOUS PREDICTOR VARIABLES

If the prognostic variable is continuous, the risk of an event would usually be expected to increase or decrease systematically as the level increases. Many researchers prefer to categorize patients by cut points. This type of analysis reduces the power to detect a real association with outcome. Many researchers are unwilling to assume that a relationship with outcome is log-linear. The assumption of linearity may will be more reasonable than the assumptions that go with dichotomize by cut points.

13.2.10 DIFFICULTIES

The prognostic studies rise several difficulties as shown below:

1. Difficulty of identifying all studies.
2. Inadequate reporting of methods.
3. Variation in study design.
4. Most studies are retrospective.
5. Variation in inclusion criteria.
6. Lack of recognized criteria for quality assessment.
7. Different assays/measurement techniques.
8. Variation in methods of analysis.
9. Differing methods of handling of continuous variables (some data dependent).
10. Different statistical methods of adjustment.
11. Adjustment for different sets of variables.
12. Inadequate reporting of quantitative information on outcomes.
13. Variation in presentation of results such as survival at different time-points.

13.2.11 STATISTICAL METHODS

The method of analysis for pooling values across several studies will depend on whether the prognostic variable is binary, categorical, or continuous. It is easy to combine data from studies which have produced compatible estimates of effect, with standard error.

13.2.12 OUTCOME IS TIME TO EVENT

In case of time to event data, the data are analyzed using survival analysis method-most often the log-rank test for simple comparisons on Cox regression for analysis of multiple predictor variables. Log-rank statistic and log-hazard ratios can be combined using the Peto method or the inverse variance method, respectively. It is possible to combine estimated Kaplan–Meier survival probabilities at a single time-point.

13.2.13 STUDIES OF MANY PROGNOSTIC FACTORS

Some meta-analysis studies consider the set of research studies where the aim was to investigate many factors simultaneously, to identify important risk factors. In this case, there is considerable risk of false-positive findings in individual studies. Pooling of quantitative estimates will be problematic.

13.3 EVALUATION OF DIAGNOSTIC TESTS META-ANALYSIS

This section focuses on systematic reviews of studies of diagnostic accuracy which describe the probabilistic relationship between positive and negative test results and the presence or absence of disease, and therefore indicate how well a test can separate diseased from non-diseased patients (Egger et al., 2001).

13.3.1 SAMPLE SIZE

Statistical power is rarely discussed in studies of diagnostic accuracy as they do not compare two groups, and they do not formally test hypotheses.

However, increasing sample size by pooling the results of several studies does improve the precision of estimates of diagnostic performance.

13.3.2   SOURCE OF HETEROGENEITY

The choice of a statistical method for pooling results depends on the source of heterogeneity, especially variation in diagnostic thresholds. Sensitivities, specificities, and likelihood ratios may be combined directly, if the results are reasonably homogeneous. When study results are strongly heterogeneous, it may be most appropriate not to attempt statistical pooling.

13.3.3   PERFORMANCE

The full evaluation of the performance of a diagnostic test involves studying test reliability, diagnostic accuracy, and therapeutic impact, and the net effect of the test on patient outcomes. Separate meta-analysis can be performed for each of these aspects of test evaluation depending on the availability of suitable studies.

13.3.4   META-ANALYSIS OF DIAGNOSTIC ACCURACY

Three general approaches used to pool results of studies of diagnostic accuracy that are described below. The selection of a method depends on the choice of a summary statistic and potential causes of heterogeneity. The choice of statistical method for combining study results depends on the pattern of heterogeneity observed between the results of the studies. Some divergence of the study results is to be expected by chance, but variation in other factors, such as patient selection and features of study design may increase the observed variability or heterogeneity.

13.3.5   DIAGNOSTIC THRESHOLD

There is also one important extra source of variation to consider in meta-analysis of diagnostic accuracy: variation introduced by changes in diagnostic threshold. The studies included in a systematic review may have used different thresholds to define positive and negative test results. Some may have done this explicitly, for example by varying numerical cut-points used to classify a biochemical measurement as a positive or negative. For others, there may be naturally occurring variations in diagnostic thresholds between observers or between laboratories. The choice of a threshold may also have been determined according to the prevalence of the disease. When the disease is a rare a low threshold may have been used to avoid large numbers of false-positive diagnoses being made. Unlike random variability and other sources of heterogeneity, varying the diagnostic threshold between studies introduces a particular pattern into the ROC plot of study results.

13.3.6 DIAGNOSTIC ODDS RATIO

Sensitivities and specificities, and positive and negative likelihood ratios can be combined into the same single summary of diagnostic performance, known as the diagnostic odds ratio. This statistic is not easy to apply in clinical practice, but it is a convenient measure to use when combining studies in a meta-analysis as it is often reasonably constant regardless of the diagnostic threshold. The diagnostic adds ratio (DOR) is defined as

$$DOR = \frac{TP \times TN}{EP \times FN}$$

where

Test results	Participants		
	With disease	**Without disease**	**Total**
Positive test	True positives (TR)	False positives (FP)	Total positive
Negative test	False negatives (FN)	True negatives (TN)	Total negative
Total	Total with disease	Total without disease	

13.3.7 POOLING DIAGNOSTIC ODDS RATIOS

If there is any evidence that the diagnostic threshold varies between the studies, the best summary of the results of the studies will be an ROC curve rather than a single point. The full method for deciding on the best-fitting summary ROC is explained below. First, it is worth noting that a simple method for estimating a summary ROC curve exists when it can be assumed that the curve is symmetrical around the "sensitivity = specificity" line.

Diagnostic tests where the diagnostic odds ratio is constant regardless of the diagnostic threshold have symmetric ROC curves. In these situations, it is possible to use standard meta-analysis methods for combining odds ratio to estimate the common diagnostic odds ratio, and hence to determine the best-fitting ROC curve. Once the summary odds ratio (DOR), has been calculated, the equation of the corresponding ROC curve is given by

$$\text{Sensitivity} = \frac{1}{1 + \left(1/DOR \times \left(1 - \text{specificity}/\text{specificity}\right)\right)}$$

13.3.8   POOLING SENSITIVITIES AND SPECIFICITIES

The simplest method of combining studies of diagnostic accuracy is to compute weighted averages of the sensitivities, specificities, or likelihood ratios. This method can only be applied in the absence of variability of the diagnostic threshold. The possibility of a threshold effect can be investigated before this method is used, both graphically by plotting the study results on an ROC plot, and statistically, by undertaking tests of the heterogeneity of sensitivities and specificities and investigating whether there is a relationship between them. The homogeneity of the sensitivities and specificities from the studies can be tested using standard chi-squared tests as both measures are simple proportions.

Computation of an average sensitivity and specificity is straight forward. Considering the sensitivity and specificity in each study i to be denoted by a proportion P_i.

$$P_i = \frac{y_i}{n_i}; \text{sensitivity}_i = \frac{\text{True positive}_i}{\text{Total with disease}_i}; \text{specificity} = \frac{\text{True negative}_i}{\text{Total without disease}_i}$$

Using an approximation to the inverse variance approach, the estimate of the overall proportion is

$$P_i = \frac{\sum y_i}{\sum n_i}$$

where Σy_i is the sum of all true positives (for sensitivity) or true negatives (for specificity), and Σn_i is the sum of diseased (for sensitivity) or not diseased (for specificity). The large sample approximation for the standard error of this estimate is

$$SE(p) = \sqrt{\frac{p(1-p)}{\sum n_i}}$$

The simplest analysis pools the estimates of sensitivity and specificity separately across the studies but is only appropriate if there is no variation in diagnostic threshold.

13.3.9   POOLING LIKELIHOOD RATIOS

Likelihood ratios are ratios of probabilities and in a meta-analysis can be treated as risk ratio. A weighted average of the likelihood ratios can be computed using the standard Mantel–Haenszel or inverse variance methods of meta-analysis of risk ratios.

KEYWORDS

- observational studies
- prognostic variables
- prediction model
- diagnostic odds ratio
- likelihood ratios

REFERENCE

Egger, M.; Smith, G. D.; Altman, D. G., Eds. *Systematic Reviews in Health Care—Meta-analysis in Context*. BMJ Publishing Group: London, 2001.

CHAPTER 14

ADDITIONAL META-ANALYSIS TECHNIQUES

CONTENTS

ABSTRACT

The sensitivity analysis is the study of influence and robustness of different statistical methodologies on the results of meta-analysis. In case of influence meta-analysis, the influence of each study can be estimated by deleting each in turn from the analysis and noting the degree to which the size and significance of the treatment effect changes. The subgroup analysis is the study of variation by different categories of patients on the results of meta-analysis. When the heterogeneity statistic is significant, it is not feasible to assume that the given treatment effect is same across different groups of patients. The subgroup analysis is carried out when there is indirect evidence suggesting considerable difference between categories. The subgroup analysis should best be used as hypothesis-generating tools. The usefulness of subgroup analysis may be limited due to small number of studies included in meta-analysis. The cumulative meta-analysis is a repeated performance of meta-analysis whenever a new relevant study is available for inclusion. This allows a retrospective identification of the patient in time whenever a treatment effect first reach conventional level of statistical significance. As in primary studies with regression, or multiple regressions, in meta-regression analysis, we assess the relationship between one or more covariates and a dependent variable. Meta-regression has become a commonly used tool for investigating whether clinical heterogeneity may explain the statistical heterogeneity. The meta-regression analysis is used to suggest reasons for statistical heterogeneity. The multiple meta-regressions can encompass two or more study characteristics simultaneously as independent variables. The main objectives of meta-cluster analysis is to find out which studies in a meta-analysis are similar and which studies dissimilar with respect to relevant variables related to the particular summary statistic of the analysis.

14.1   SENSITIVITY META-ANALYSIS

The sensitivity analysis is the study of influence and robustness of different statistical methodologies on the results of meta-analysis. It is also a repeat of the primary analysis, substituting alternative decisions or ranges of values for decisions that were arbitrary or unclear. The meta-analysis for different methodological issues of primary studies such as published versus unpublished studies, randomization versus non-randomization, different levels of quality assessment, different sample sizes, and fixed-effects over random-effects models have to be examined for variation. Such analysis in a graphical display allows for visual examination of the sensitivity of important methodologies adopted in the primary studies.

14.1.1   USE OF SENSITIVITY META-ANALYSIS

While conducting meta-analysis researchers are bound to work on certain assumptions like inclusion exclusion criteria, quality of the study, choice of fixed or random-effects model, etc. There is diverging opinions on the correct method for performing a particular meta-analysis. The robustness of the findings to different assumptions should therefore always be examined. This can be achieved through sensitivity analysis.

Heterogeneity is typically explored using sensitivity analysis where associations are estimated under various assumptions. If the sensitivity analyses that are done do not materially change the results, it strengthens the confidence that can be placed in them. If the results change in a way that might lead to different conclusions, this indicates a need for greater caution in interpreting the results and drawing conclusions, or it may generate hypothesis for further investigations (Tharyan, 1998).

14.1.2   GRAPHICAL PRESENTATION

The graphical presentation of sensitivity meta-analysis based on several assumptions made while studying a pattern and prevalence of epilepsy in India based on 47 selected studies is presented in the form of Forest plot in Figure 14.1.

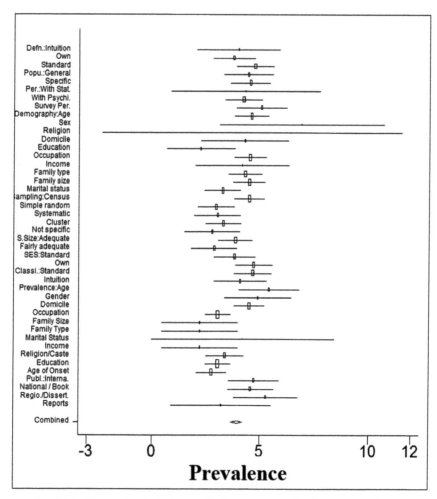

FIGURE 14.1   Forest plot depicting sensitivity meta-analysis of epilepsy studies.

14.2   INFLUENCE META-ANALYSIS

The influence of each study can be estimated by deleting each in turn from
the analysis and noting the degree to which the size and significance of
the treatment effect changes. The influence of individual studies on the
summary effect estimate may be displayed.

14.2.1   GRAPHICAL PRESENTATION

The graphical presentation of influence analysis made while studying a pattern and prevalence of epilepsy in India based on 47 selected studies is presented in Figure 14.2.

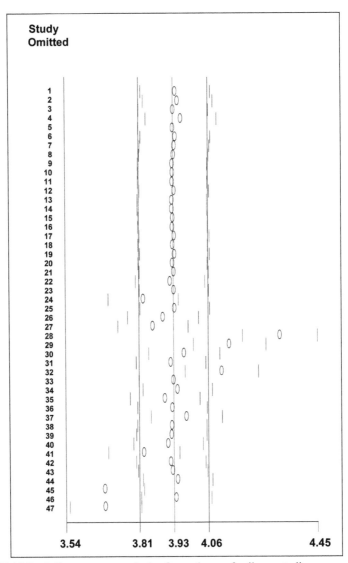

FIGURE 14.2   Influence meta-analysis of prevalence of epilepsy studies.

14.3   SUBGROUP META-ANALYSIS

The subgroup analysis is the study of variation by different categories of patients on the results of meta-analysis. It is more commonly linked to covariate analysis in statistical literature. The estimate of the overall effect is intended to guide decision about clinical practice for a wide range of patients.

14.3.1   TREATMENT DECISION

When the heterogeneity statistic is significant, it is not feasible to assume that the given treatment effect is same across different groups of patients, such as males versus females, young versus elderly, those with mild versus those with severe diseases. It seems reasonable to base treatment decision upon the results of those studies, which have similar characteristics to the particular patient under consideration.

14.3.2   CAUSE OF HETEROGENEITY

The subgroup analysis is carried out when there is indirect evidence suggesting considerable difference between categories, the difference is suggested by comparisons within studies rather than between studies, the difference is consistent across studies, and the magnitude of difference is practically important (Tharyan, 1998). If the studies within subgroup are relatively homogeneous but there is considerable between subgroup variations, an important cause of heterogeneity may be identified.

14.3.3   HYPOTHESIS-GENERATING TEST

The subgroup analysis should best be used as hypothesis-generating tools, although important observations may sometimes be made (Lau et al., 1997).

14.3.4   LIMITATIONS

The usefulness of subgroup analysis may be limited due to small number of studies included in meta-analysis. It is often not convenient to consider simultaneously more than one variable in a subgroup analysis. Subgroup analysis may be considered exploratory or confirmatory. Combining specific subgroup data across studies may provide further insight into heterogeneity. Lack of uniform reporting of the data necessarily for subgroup analyses across studies poses an additional problem. An especially pernicious approach occurs when the data are divided into multiple subgroups on the basis of combination of characteristics such as age, sex, and domicile, and different prevalence rates are claimed within very small subdivisions. Such interventions among subgroups are unlikely to describe the truth when derived from aggregate data.

14.3.5 ILLUSTRATION

The results of subgroup meta-analysis conducted based on 16 categories while studying a pattern and prevalence of epilepsy in India based on 47 selected studies is presented in the following table:

Study No.	Characteristics (number of studies)	Prevalence rate	Study No.	Characteristics (number of studies)	Prevalence rate
1	**Domicile (47)**		13	**Domicile & sex (22)**	
	Rural	4.99		*Rural* Male	4.84
	Urban	4.29		Female	4.00
	Semi-urban/Mixed	3.64		*Urban* Male	5.23
2	**Region (47)**			Female	3.41
	Northern	3.67	14	**Caste & sex (3)**	
	Eastern	4.55		*Brahmins* Male	4.60
	Western	3.32		Female	1.68
	Southern	5.21		*SC* Male	3.22
3	**Sex (22)**			Female	3.12
	Male			*ST* Male	4.04
	Female			Female	1.90
4	**Age (8)**		15	**Age & sex (1)**	
	0–	4.40		*0–* Male	4.37
	1–	7.19		Female	3.17
	5–	9.39		*5–* Male	9.54
	10–	7.61		Female	9.23
	20–	6.72		*10–* Male	7.23
	30–	5.17		Female	5.20

TABLE *(Continued)*

Study No.	Characteristics (number of studies)	Prevalence rate	Study No.	Characteristics (number of studies)	Prevalence rate
	40–	4.50		*20–* Male	7.04
	50–	3.31		Female	6.56
	60–	2.79		*30–* Male	4.19
	70–	1.82		Female	4.90
5	**Marital status (2)**				
	Single	8.61		*40–* Male	3.85
	Married	1.28		Female	4.51
6	**Religion (2)**			*50–* Male	3.26
	Hindu	2.59		Female	3.37
	Muslim	3.50		*60–* Male	2.44
7	**Caste group (5)**			Female	3.11
	Brahmins	2.85	16	**Age & domicile (1)**	
	SC	2.77		*0–9 age*	
	ST	3.48		Rural	8.10
	All others	6.93		Urban	6.92
8	**Literacy level (1)**				
	Illiterate/Primary	2.38		*0–19 age*	
	Secondary	–		Rural	16.43
	University	–		Urban	10.65
9	**Occupation (1)**			*20–29 age*	
	Children	4.29		Rural	9.81
	Students	4.51		Urban	11.27
	House-wives	1.41		*30–39 age*	
	All others	–		Rural	10.31
10	**Monthly income (1)**			Urban	7.84
	Low	6.50		*40–49 age*	
	Middle	1.04		Rural	8.93
	High	–		Urban	2.66
11	**Family type (1)**			*49+ age*	
	Nuclear	–		Rural	3.31
	Joint	3.65		Urban	2.03
12	**Family size (1)**				
	Up to 5	–			
	Above 5	5.59			

14.4   CUMULATIVE META-ANALYSIS

The cumulative meta-analysis is a repeated performance of meta-analysis whenever a new relevant study is available for inclusion. This allows a retrospective identification of the patient in time whenever a treatment effect first reached conventional level of statistical significance. Subsequent studies will simply confirm the original result with reduced levels of significance. The further studies in large number of patients involving high cost may be avoided by conducting timely cumulative meta-analysis (Hafner, 1987).

14.4.1   ADVANTAGES

Another application of cumulative meta-analysis has been to correlate the accruing evidence with the recommendations made by experts in review articles and text books (Egger et al., 2001). Cumulative meta-analysis is another approach for assessing the impact of each study. The prior probability is generated by the pooled results of all prior studies, and the posterior probability is derived by adding the results of the new study to the results of the others (Lau et al., 1992, 1995).

The cumulative plot is a special use of forest plot for the purpose of cumulative meta-analysis according to the given criteria. It can also be used to examine the influence of individual studies on the pooled estimate of treatment effect according to other study features such as study quality assessment score, control group event rate, and sample size (Lau et al., 1995).

14.4.2   GRAPHICAL REPRESENTATION

The graphical presentation of cumulative meta-analysis conducted while studying a pattern and prevalence of epilepsy in India based on 47 selected studies is presented in Figure 14.3.

14.5   META-REGRESSION ANALYSIS

As in primary studies with regression, or multiple regressions, we assess the relationship between one or more covariates and a dependent variable. Essentially the same approach can be used with meta-analysis, except that

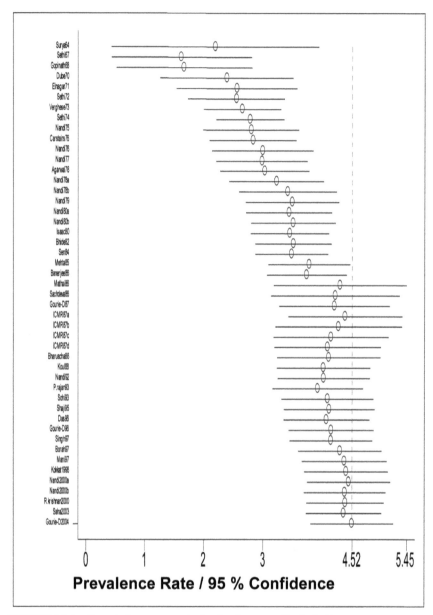

FIGURE 14.3   Cumulative meta-analysis of epilepsy studies in India.

the covariates are at the level of the study rather than the level of the subject, and the dependent variable is the effect size in the studies rather than subject scores. We use the term meta-regression to refer to these procedures when they are used in a meta-analysis.

Meta-regression has become a commonly used tool for investigating whether clinical heterogeneity may explain the statistical heterogeneity. If there is a linear relationship between treatment effect and identified study characteristics, regression analysis will have a greater statistical power than subgroup analysis.

14.5.1 SIMILARITIES WITH SUBGROUP ANALYSIS

The differences that we need to address as we move from primary studies to meta-analysis for regression are similar to those we needed to address as we moved from primary studies to meta-analysis for subgroup analyses. These include the need to assign a weight to each study and the need to select the appropriate model. Also, as was true for subgroup analyses, which is used to quantify the proportion of variance explained by the covariates, must be modified for use in meta-analysis.

14.5.2 USE OF COVARIATES

We can work with sets of covariates, such as three variables that together define a treatment, or that allow for a nonlinear relationship between covariates and the effect size. We can enter covariates into the analysis using a predefined sequence and assess the impact of any set, over and above the impact of prior sets, to control for confounding variables. We can incorporate both categorical and continuous variables as covariates.

14.5.3 LIMITATIONS

The use of meta-regression, especially with multiple covariates, is not a recommended option when the number of studies is small. In primary studies, some have recommended a ratio of at least 10 subjects for each covariate, which would correspond to 10 studies for each covariate in meta-regression.

14.5.4   *GENERATE NEW HYPOTHESIS*

The meta-regression analysis is used to suggest reasons for statistical heterogeneity. As an exploratory tool, meta-regression may provide more insight than traditional pooling and may help to generate new hypotheses (Berkey et al., 1995).

14.5.5   *COLLINEARITY*

The multiple meta-regressions can encompass two or more study characteristics simultaneously as independent variables. However, if exploratory variables are highly correlated, collinearity may become a problem. When collinearity is a problem, the interpretations of the estimated occurrence of confounding variables should be cautious or redundant variables are removed from the model. A variety of statistical methods, including weighted least squares, logistic regression, and hierarchical models can be used for meta-regression analysis (Lau et al., 1997). The random-effects meta-regression model allows the inclusion of study specific covariates, which may explain a part of the heterogeneity (Knapp and Hartung, 2003).

14.5.4   *LIMITATIONS*

Lau et al. (1998) have shown that fixed effects meta-regression is likely to produce seriously misleading results in the presence of heterogeneity.

1. It is unwise to include large number of covariates particularly if the sample size is small.
2. All associations noticed in analysis are observational and may therefore be confounded by other unknown or unmeasured factors.
3. Regression analysis using averages of patient characteristics from each trial can give a misleading impression of the relation for individual patients.

Standard meta-regression methods suffer from substantially inflated false-positive rates when heterogeneity is present, when there are few studies, and when there are many covariates (Higgins and Thompson, 2004).

14.5.6   GRAPHICAL REPRESENTATION

The graphical presentation of two meta-regression analyses conducted while studying a pattern and prevalence of epilepsy in India based on 47 selected studies taking year of report and quality assessment score as independent variables are presented in Figures 14.4 and 14.5.

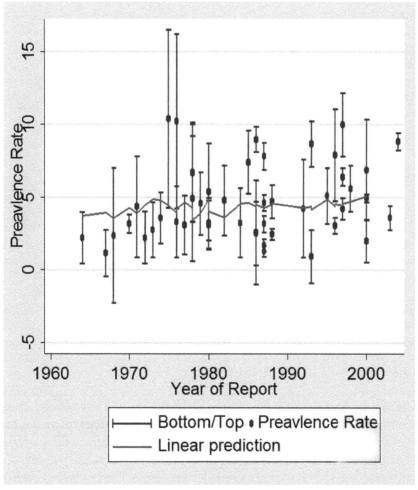

FIGURE 14.4   Meta-regression plot of prevalence rates on year of report for epilepsy.

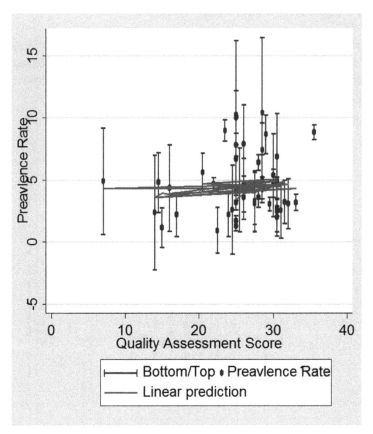

FIGURE 14.5 Meta-regression plot of prevalence on quality assessment scores for epilepsy.

14.6 META-CLUSTER ANALYSIS

The main objectives of meta-cluster analysis is to find out which studies in a meta-analysis are similar and which studies dissimilar with respect to relevant variables related to the particular summary statistic of the analysis. The relevant variables include the quality assessment scores and psychosocial variables. The clustering techniques may be used to produce groups of studies with different pooled estimates. Thus, it enables the investigator to determine the overall pooled estimate in a heterogeneous population. The studies are summarized by referring to the properties of the cluster rather than the properties of the individual studies. In addition to the graphical methods, random-effects model of pooled estimates, sensitivity and influential analysis, cumulative analysis, and meta-regression analysis, cluster meta-analysis is another technique to study the inter-study variation.

14.6.1 GRAPHICAL REPRESENTATION

The graphical presentation of meta-cluster analyses conducted while studying a pattern and prevalence of epilepsy in India based on 47 selected studies is presented in Figure 14.6.

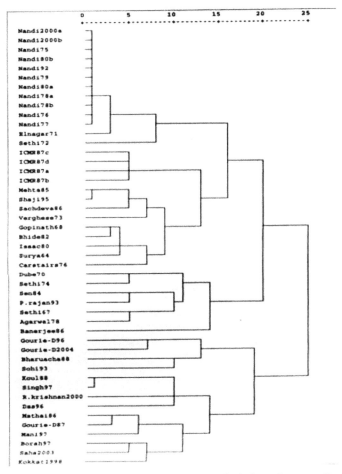

FIGURE 14.6 Dendrogram depicting cluster meta-analysis for epilepsy studies in India.

KEYWORDS

- sensitivity analysis
- influence meta-analysis
- subgroup analysis
- cumulative meta-analysis
- meta-regression analysis
- meta-cluster analysis

REFERENCES

Agarwal, R. B. Epidemiological Study of Psychiatric Disorders in Urban Setting (200 Families). M. D. Thesis, Gujarat University: Gujarat, 1978; pp 37–38.

Banerjee, T.; Mukherjee, S. P.; Nandi, D. N.; Banerjee, G.; Mukherjee, A.; Sen, B.; Sarker, G.; Boral, G. C. Psychiatric Morbidity in an Urbanized Tribal (Santal) Community—A Field Survey. *Indian J. Psychiatry* **1986**, *28*, 243–248.

Berkey, C. S.; Hoaglin, D. C.; Mosteller, F.; Colditz, G. A. A Random Effects Regression Model for Meta-analysis. *Stat. Med.* **1995**, *14*, 395–411.

Bharucha, N. E.; Bharucha, E. P.; Bharucha, A. E.; Bhise, A. V.; Schoenberg, B. S. Prevalence of Epilepsy in the Parsi Community of Bombay. *Epilepsia* **1988**, *29*, 111–115.

Bhide, A. Prevalence of Psychiatric Morbidity in a closed community in South India, M.D. Thesis, National Institute of Mental Health and Neurosciences: Bangalore, 1982.

Borah. *Medical Times*; Sandoz Publication: Mumbai, 1997; Vol 27, pp 1–3.

Carstairs, G. M.; Kapur, R. L. The Great Universe of Kota—Stress Change and Mental Disorder in an Indian Village. Hogarth Press: London, **1976**; pp 89–114.

Das, S. K.; Sanyal, K. Neuroepidemiology of Major Neurological Disorders in Rural Bengal. *Neurol. India* **1996**, *44*, 47–58.

Dube, K. C. A Study of Prevalence and Biosocial Variables in Mental Illness in a Rural and an Urban Community in Uttar Pradesh, India. *Acta Psychiatr. Scand.* **1970**, *46*, 327–359.

Egger, M.; Smith, G. D.; Altman, D. G., Eds. Systematic Reviews in Health Care—Meta-analysis in Context. BMJ Publishing Group: London, 2001.

Elnagar, M. N.; Maitra, P.; Rao, M. N. Mental Health in an Indian Rural Community. *Br. J. Psychiatry* **1971**, *118*, 499–503.

Gopinath, P. S. Epidemiology of Mental Illness in an Indian Village, M.D. Thesis, Bangalore University: Bangalore, 1968; pp 54–71.

Gourie-Devi, M.; Gururaj, G.; Satishchandra, P.; Subbakrishna, D. K. Neuro-epidemiological Pilot Survey of an Urban Population in a Developing Country. *Neuroepidemiology* **1996**, *15*, 313–320.

Gourie-Devi, M.; Gururaj, G.; Satishchandra, P.; Subbakrishna, D. K. Prevalence of Neurological Disorders in Bangalore, India: A Community Based Study with a Comparison between Urban and Rural Areas. *Neuroepidemiology* **2004**, *23*, 261–268.

Gourie-Devi, M.; Rao, V. N.; Prakashi, R. Neuroepidemiological Study in Semiurban and Rural Areas in South India: Pattern of Neurological Disorders Including Motor Neurone Disease. In *Motor Neuron Diseases*; Gourie-Devi, M., Ed.; Oxford and IBH Publication: New Delhi, 1987; pp 11–21.

Hafner, H. In *Search for the Causes of Schizophrenia*; Hafner, H., Gattaz, W. F., Janjarik, W., Eds.; Springer Verlag: Berlin, Heidelberg, New York, 1987.

Higgins, J. P. T.; Thompson, S. G. Controlling the Risk of Spurious Findings from Meta-regression. *Stat. Med.* **2004**, *23*, 1663–1682.

Indian Council of Medical Research Report. In *Collaborative Study on Severe Mental Morbidity*; Issac, M. K., Ed.; Indian Council of Medical Research and Department of Science and Technology: New Delhi, 1987a,b,c,d.

Issac, M. K.; Kapur, R. L. A Cost-Effectiveness Analysis of Three Different Methods of Psychiatric Case Finding in the General Population. *Br. J. Psychiatry* **1980**, *137*, 540–546.

Knapp, G.; Hartung, J. Improved Tests for a Random Effects Meta-regression with a Single Covariate. *Stat. Med.* **2003**, *22*, 2693–2710.

Kokkat, A. J.; Verma, A. K. Prevalence of Seizures and Paralysis in a Rural Community. *J. Indian Med. Assoc.* **1998**, *96*, 43–45.

Koul, R.; Razdan, S.; Motta, A. Prevalence and Pattern of Epilepsy (Lath/Mirgi/Laran) in Rural Kashmir, India. *Epilepsia* **1988**, *29*, 116–122.

Lau, J.; Loannidis, J. P. A.; Schmid, C. H. Quantitative Synthesis in Systematic Reviews. *Ann. Intern. Med.* **1997**, *127*, 820–826.

Lau, J.; Loannidis, J. P.; Schmid, C. H. Summing Up Evidence: One Answer is Not Always Enough. *Lancet* **1998**, *351*, 123–127.

Lau, J.; Schmid, C.; Chalmers, T. C. Cumulative Meta-analysis of Clinical Trials Builds Evidence for Exemplary Medical Care. *J. Clin. Epidemiol.* **1995**, *48*, 45–57.

Lau. J.; Antman, E. M.; Jimenez-Silva, J.; Kupenick, B. M. D.; Mosteller, B. A. F.; Chalmers, T. C. Cumulative Meta-analysis of Therapeutic Trials for Myocardial Infarction. *N. Engl. J. Med.* **1992**, *327*, 248–254.

Mani, K. S.; Rangan, G.; Srinivas, H. V.; Kalyanasundaram, S.; Narendran, S.; Reddy, A. K. The Yelandur Study: A Community Based Approach to Epilepsy in Rural South India—Epidemiological Aspects. *Seizure* **1997**, *7*, 281–288.

Mathai, K. V. Epilepsy—Some Epidemiological, Experimental and Surgical Aspects. *Neurol. India* **1986**, *34*, 299–314.

Mehta, P.; Joseph, A.; Verghese, A. An Epidemiological Study of Psychiatric Disorders in a Rural Area in Tamilnadu. *Indian J. Psychiatry* **1985**, *27*, 153–158.

Nandi, D. N.; Ajmany, S.; Ganguli, H.; Banerjee, G.; Boral, G. C.; Ghosh, A.; Sarkar, S. The Incidence of Mental Disorders in One Year in a Rural Community in West Bengal. *Indian J. Psychiatry* **1976**, *18*, 79–87.

Nandi, D. N.; Ajmany, S.; Ganguly, H.; Banerjee, G.; Boral, G. C.; Ghosh, A. Psychiatric Disorders in a Rural Community in West Bengal: An Epidemiological Study. *Indian J. Psychiatry* **1975**, *17*, 87–99.

Nandi, D. N.; Banerjee, G.; Boral, G. C.; Ganguli, H.; Ajmany, S.; Ghosh, A.; Sarkar, S. Socio-Economic Status and Prevalence of Mental Disorders in Certain Rural Communities in India. *Acta Psychiatr. Scand.* **1979,** *59,* 276–293.

Nandi, D. N.; Banerjee, G.; Chowdhury, A. N.; Banerjee, T.; Boral, G. C.; Sen, B. Urbanisation and Mental Morbidity in Certain Tribal Communities in West Bengal. *Indian J. Psychiatry* **1992,** *34,* 334–339.

Nandi, D. N.; Banerjee, G.; Ganguli, H.; Ajmany, S.; Boral, G. C.; Ghosh, A.; Sarkar, S. The Natural History of Mental Disorders in a Rural Community: A Longitudinal Field Survey. *Indian J. Psychiatry* **1978b,** *21,* 390–396.

Nandi, D. N.; Banerjee, G.; Mukherjee, S. P.; Ghosh, A.; Nandi, P. S.; Nandi, S. Psychiatric Morbidity of a Rural Indian Community—Changes Over a 20-Year Interval. *Br. J. Psychiatry* **2000a,b,** *176,* 351–356.

Nandi, D. N.; Das, N. N.; Chaudhuri, A.; Banerjee, G.; Datta, P.; Ghosh, A.; Boral, G. C. Mental Morbidity and Urban Life: An Epidemiological Study. *Indian J. Psychiatry* **1980b,** *22,* 324–330.

Nandi, D. N.; Mukherjee, S. P.; Banerjee, G.; Boral, G. C.; Ghosh, A.; Sarkar, S.; Das, S.; Banerjee, K.; Ajmany, S. Psychiatric Morbidity in an Uprooted Community in Rural West Bengal. *Indian J. Psychiatry* **1978a,** *20,* 137–142.

Nandi, D. N.; Mukherjee, S. P.; Boral, G. C.; Banerjee, G.; Ghosh, A.; Ajmany, S.; Sarkar, S.; Biswas, D. Prevalence of Psychiatric Morbidity in Two Tribal Communities in Certain Villages of West Bengal: A Cross Cultural Study. *Indian J. Psychiatry* **1977,** *19,* 2–12.

Nandi, D. N.; Mukherjee, S. P.; Boral, G. C.; Banerjee, G.; Ghosh, A.; Sarkar, S.; Ajmany, S. Socio-economic Status and Mental Morbidity in Certain Tribes and Castes in India: A Cross-Cultural Study. *Br. J. Psychiatry* **1980a,** *136,* 73–85.

Premarajan, K. C.; Danabalan, M.; Chandrasekar, R.; Srinivas, D. K. Prevalence of Psychiatric Morbidity in an Urban Community of Pondicherry. *Indian J. Psychiatry* **1993,** *35,* 99–102.

Radhakrishnan, K.; Pandian, J. D.; Santhoshkumar, T.; Thomas, S. V.; Deetha, T. D.; Sarma, P. S.; Jayachandran, D.; Mohamed, E. Prevalence, Knowledge, Attitude, and Practice of Epilepsy in Kerala, South India. *Epilepsia* **2000,** *41,* 1027–1035.

Sachdeva, J. S.; Singh, S.; Sidhu, B. S.; Goyal, R. K. D.; Singh, J. An Epidemiological Study of Psychiatric Disorders in Rural Faridkot (Punjab). *Indian J. Psychiatry* **1986,** *28,* 317–323.

Saha, S. P.; Bhattacharya, S.; Das, S. K.; Maity, B.; Roy, T.; Raut, D. K. Epidemiological Study of Neurological Disorders in a Rural Population of Eastern India. *J. Indian Med. Assoc.* **2003,** *101,* 299–300.

Sen, B.; Nandi, D. N.; Mukherjee, S. P.; Mishra, D. C.; Banerjee, G.; Sarkar, S. Psychiatric Morbidity in an Urban Slum-dwelling Community. *Indian J. Psychiatry* **1984,** *26,* 185–193.

Sethi, B. B.; Gupta, S. C.; Mahendru, R. K.; Kumari, P. Mental Health and Urban Life: A Study of 850 Families. *Br. J. Psychiatry* **1974,** *124,* 243–246.

Sethi, B. B.; Gupta, S. C.; Rajkumar, Kumari, P. A Psychiatric Survey of 500 Rural Families. *Indian J. Psychiatry* **1972,** *14,* 183–196.

Sethi, B. B.; Gupta, S. C.; Rajkumar. Three Hundred Urban Families: A Psychiatric Study. *Indian J. Psychiatry* **1967,** *9,* 280–302.

Shaji, S.; Verghese, A.; Promodu, K.; George, B.; Shibu, V. P. Prevalence of Priority Psychiatric Disorders in a Rural Area in Kerala. *Indian J. Psychiatry* **1995**, *37*, 91–96.

Singh, A.; Kaur, A. Epilepsy in Rural Haryana—Prevalence and Treatment Seeking Bahaviour. *J. Indian Med. Assoc.* **1997**, *95*, 37–39.

Sohi, D.; Walia, I.; Singh, A. Prevalence and Treatment of Epilepsy in a Chandigarh Slum. *Bull. PGI, Chandigarh* **1993**, *27*, 175–178.

Surya, N. C.; Datta, S. P.; Gopalakrishna, R.; Sundaram, D.; Kutty, J. Mental Morbidity in Pondicherry (1962–1963). *Trans. All Indian Inst. Ment. Health (NIMHANS)* **1964**, *4*, 50–61.

Tharyan, P. The Relevance of Meta-analysis, Systematic Reviews and the Cochrane Collaboration to Clinical Psychiatry. *Indian J. Psychiatry* **1998**, *40*, 135–148.

Verghese, A.; Beig, A.; Senseman, L. A.; Rao, S. S. S.; Benjimin, V. A Social and Psychiatric Study of a Representative Group of Families in Vellore Town. *Indian J. Med. Res.* **1973**, *61*, 608–620.

CHAPTER 15

REPORTING META-ANALYSIS RESULTS

CONTENTS

ABSTRACT

The meta-analysis should contain enough studies to provide power for its test. If a meta-analysis performs moderator tests, it should also report if there are any relationships between the moderators. All meta-analyses will have at least two authors to ensure coding reliability. If there are a large number of assumed zero effect sizes, the authors should report their results both including and excluding these values from their analyses. To assess the representativeness of a particular meta-analysis one should consider theoretical boundaries, exhaustive search, secondary literature, unpublished literature, large literature, and high inference moderation. A meta-analysis should not simply be a summary of a literature but should provide a theoretical interpretation and integration. A good meta-analysis puts effort into interpreting these findings, presents how they are consistent or inconsistent with the major theories in the literature and encourages future investigations. Although it can be argued that the results of a systematic review should stand on their own, many people faced with a decision look to the discussion and authors' conclusions for interpreting the results. Indeed, many people prefer to go directly to the conclusions before looking at the rest of the review. The discussion and conclusions should help people to understand the implications of the evidence in relationship to practical decisions. A good starting point for the discussion section of a review is to address any important methodological limitations of the included trials and the methods used in the review that might affect practical decisions about healthcare or future research. One type of evidence that can be helpful in considering the likelihood of a cause–effect relationship between an intervention and an important outcome is indirect evidence of a relationship. Decisions about applicability depend on knowledge of the particular circumstances in which decisions about healthcare are being made. Important variations include biologic, cultural, compliance, baseline risk, and results of the included studies. It is safer to report the data, with a confidence interval, as being compatible with either a reduction or an increase in the outcome. The easier way to write up a meta-analysis is to take advantage of this parallel structure by using the same sections found in primary research. In the method section, you need to describe how you collected your studies and how you obtained quantitative codes. In the results section, you describe the distribution of your effect sizes and present any moderator analyses you decided to perform.

Discussion and conclusion section conclude with specific recommendations for the direction of future research. Reference and appendix section have a single reference section that includes both studies used in writing the paper and those included in the meta-analysis.

15.1 EVALUATION OF META-ANALYSIS STUDY

15.1.1 INTERNAL VALIDITY

Primary studies: A meta-analysis can never be more valid than the primary studies that it is aggregating. If there are methodological problems with the studies, then the validity of the meta-analysis should be equally called into question.

Power of the test: The meta-analysis should contain enough studies to provide power for its test. The exact number will depend on what analyses are being performed. For most purposes, you would want to have at least 30 studies are needed.

Correlation between moderators: If a meta-analysis performs moderator tests, it should also report if there are any relationships between the moderators. We should critically examine all results involving correlated moderators to see if there is a logical reason to doubt the interpretation of the results.

Coding reliability: All meta-analyses will have at least two authors to ensure coding reliability. The reliability should be published and should be reasonably high, preferably over 0.8.

Assumptions: Standard meta-analytic procedures assume that all of the effect sizes are independent. If an analysis includes more than a single effect size per study, this assumption is violated. Sometimes the designs of the primary studies will require this violation, but the authors should take steps to minimize its impact on their results.

Zero effect size: Assumed zero effect sizes from reported null findings are the least precise effects that can be calculated. We should be cautious when drawing inferences from a meta-analysis that contains a substantial amount of these effects. If there are a large number of assumed zero effect sizes, the authors should report their results both including and excluding these values from their analyses.

15.1.2 EXTERNAL VALIDITY

The most important factor affecting the external validity of a meta-analysis is the representativeness of the sample of studies (DeCoster, 2009). Ideally, the sample of a meta-analysis should contain every study that has been conducted bearing on the topic of interest. To assess the representativeness of a particular meta-analysis, one should consider the following:

Theoretical boundaries: Theoretical boundaries proposed by the authors make sense. The studies in the analysis actually compose a literature unto themselves. Sometimes they can be too broad, such that they aggregate dissimilar studies. Other times they may be too narrow, such that the scope of the meta-analysis is smaller than the scope of the theories developed in the area.

The effects calculated for each study should represent the same theoretical construct. While the specifics may be dependent on the study methodology, they should all clearly be examples of the same concept.

Exhaustive search: The authors conduct a truly exhaustive literature search. You should evaluate the keywords they used in their computer searches, and what methods they used to locate studies other than computer searches.

Secondary literature: While the majority of the studies will likely come from a single literature, it is important to consider what other fields might have conducted research related to the topic.

Unpublished: If the authors include unpublished articles, they must know how rigorous was the search. If they did not, do they provide a justification for this decision?

Large literature: If there are too many studies to reasonably include them all in the analysis, a random sample should be selected from the total population.

High inference moderation: If the analysis included high-inference coding, the report should state the specifics of how this was performed and what steps they took to ensure validity and reliability. All high-inference moderators deserve to be looked at closely and carefully.

15.1.3 THEORETICAL CONTRIBUTION

Theoretical interpretation: A meta-analysis should not simply be a summary of a literature, but should provide a theoretical interpretation and integration. In general, the more a meta-analysis provides beyond its statistical calculations, the more valuable its scientific contribution.

Consistency of findings: A good meta-analysis does not simply report main effect and moderator tests. It also puts effort into interpreting these findings and presents how they are consistent or inconsistent with the major theories in the literature.

Gaps in the area of research: Meta-analyses can greatly aid a literature by providing a retrospective summary of what can be found in the existing literature. This should be followed by suggestions of what areas within the literature still need development. A good meta-analysis encourages rather than impedes future investigations.

15.2   INTERPRETING THE RESULTS

When interpreting the results, reviews should consider the importance of beneficial and harmful effects of intervention in absolute and relative terms and address economic implications for future research.

15.2.1   ISSUES IN INTERPRETATION

Although it can be argued that the results of a systematic review should stand on their own, many people faced with a decision look to the Discussion and Authors' Conclusions for help interpreting the results. Indeed, many people prefer to go directly to the conclusions before looking at the rest of the review.

Discussion and conclusions about the following issues can help people to make decisions:

- The strength of the evidence.
- The applicability of the results.
- Considerations of costs and current practice that might be relevant to someone making a decision.
- Clarification of any important trade-offs between the expected benefits, harms, and costs of the intervention.

Authors should be particularly careful to bear in mind that different people might make different decisions based on the same evidence. The primary purpose of the review should be to present information, rather than to offer advice. The discussion and conclusions should help people to understand the implications of the evidence in relationship to practical decisions. Recommendations that depend on assumptions about resources and values should be avoided.

15.2.2   STRENGTH OF EVIDENCE

A good starting point for the discussion section of a review is to address any important methodological limitations of the included trials and the methods used in the review that might affect practical decisions about health care or future research. It is often helpful to discuss how the included studies fit into the context of other evidence that is not included in the review.

Cause and effect relationship: One type of evidence that can be helpful in considering the likelihood of a cause–effect relationship between an intervention and an important outcome is indirect evidence of a relationship.

Because conclusions regarding the strength of inferences about the effectiveness of an intervention are essentially causal inferences, authors might want to consider guidelines for assessing the strength of a causal inference. In the context of a systematic review of clinical trials, these considerations might include:

- quality of the included trials,
- significance of the observed effects,
- consistency are the effects across trials,
- clear dose-response relationship, and
- indirect evidence that supports the influence.

15.2.3   APPLICABILITY

Decisions about applicability depend on knowledge of the particular circumstances in which decisions about health care are being made. In addressing the applicability of the results of a review, authors should be cautious not to assume that their own circumstances, or the circumstances reflected in the included studies are necessarily the same as those of others. Generally, rather than rigidly applying the inclusion and exclusion criteria of studies to specific circumstances, it is better to ask whether there are compelling reasons why the evidence should not be applied under certain circumstances.

Authors can sometimes help people making specific decisions by identifying important variation where divergence might limit the applicability of results.

15.2.4 IMPORTANT VARIATIONS

Biologic and cultural variation: Issues of biologic variation that might be considered include divergence in pathophysiology and divergence in a causative agent. For some health-care problems, such as psychiatric problems, cultural difference can sometimes limit the applicability of results.

Variation in compliance: Variation in the compliance of the recipient and providers of care can limit the applicability of results. Predictable difference in compliance can be due to divergence in economic conditions or attitudes that make some form of care not accessible or not feasible in some settings, such as in developing countries.

Variation in baseline risk: The net benefit of any intervention depends on the risk of adverse outcomes without intervention, as well as on the effectiveness of the intervention. Therefore, variation in baseline risk is almost always an important consideration in determining the applicability of results.

Variation in the results of the included studies: In addition to identifying limitations of the applicability of the results of their review, authors should discuss and draw conclusions about important variation in results within the circumstances to which the results are applicable. Is there predictable variation in the relative effects of the intervention and are there identifiable factors that may cause the response or effects to vary. These might include:

- Patient features, such as age, sex, biochemical markers
- Intervention features, such as the timing or intensity of the intervention
- Disease features, such as hormone receptor status

These features should be examined even if there is not statistically significant heterogeneity. This should be done by testing whether there is an interaction with treatment, and not by subgroup analysis. As discussed, differences between subgroups, particularly those that correspond to differences between studies, need to be interpreted cautiously. Some chance variation between subgroups is inevitable, so unless there is strong evidence of an interaction then it should be assumed there is none.

Other relevant information: It can be helpful for authors to discuss the results of a review in the context of other relevant information, such as epidemiological data about the magnitude and distribution of a problem, information about current clinical practice from administrative databases or practice surveys, and information about costs.

Adverse effects: The discussion and conclusions of a review should note the strength of the evidence on adverse effects including the estimates of their seriousness and frequency in different circumstances. In partic-ular, the causal relationship of an adverse effect to a particular intervention should be critically assessed, bearing in mind that under-ascertainment and under-reporting of adverse and unexpected effects are common.

Trade-offs: Health-care interventions generally entail costs and risks of harm, as well as expectations of benefit. Drawing conclusions about the practical usefulness of an intervention entails making trade-offs, either implicitly or explicitly, between the estimated benefits and the estimated costs and harms.

Common errors in reaching conclusions: It is safer to report the data, with a confidence interval, as being compatible with either a reduction or an increase in the outcome. When there is positive but statistically nonsig-nificant trend authors commonly describe this as "promising," whereas a "negative" effect of the same magnitude is not commonly described as a "warning sign." Authors should be careful not to do this. Another mistake is to frame the conclusion in wishful terms.

Another common mistake is to reach conclusions that go beyond the evidence that is reviewed. Often this is done implicitly, without refer-ring to the additional information or judgments that are used in reaching conclusions about the implications of a review for practice. Even when conclusions about the implications of a review for practice are supported by additional information and explicit judgments, the additional informa-tion that is considered is rarely systematically reviewed and implications for practice are often dependent on specific circumstances and values that must be taken into consideration.

15.3   WRITING THE REPORT

The easier way to write up a meta-analysis is to take advantage of this parallel structure by using the same sections found in primary research. When writing a quantitative literature review, you should therefore include sections for the introduction, methods, results, and discussion.

If your summary includes moderator analyses, you should present it as a separate study in your paper, using the guidelines for reporting a quantitative review. However, if you are only presenting descriptive analyses, your meta-analysis will likely be simple enough that you can incorporate it directly into your introduction or discussion. In this case, you should describe the purpose and method of your meta-analysis in one paragraph, with the results and discussion in a second.

Your introduction should concretely define the topic of your analysis and place that topic into a broader psychological context.

15.3.1   METHOD SECTION

In the method section, you need to describe how you collected your studies and how you obtained quantitative codes from them:

(1) to describe how you collected your studies,
(2) a thorough description of your search procedure,
(3) to describe how you coded moderator variables,
(4) descriptions of each moderator you coded, and
(5) to describe how you calculated your effect sizes.

15.3.2   RESULTS SECTION

In the results section, you describe the distribution of your effect sizes and present any moderator analyses you decided to perform:

• to describe the distribution of your effect sizes,
• to categorize moderator you want to test,
• to test continuous variable, and
• to describe the ability of a statistical model to explain your distribution of effect sizes.

15.3.3 DISCUSSION AND CONCLUSION SECTION

To help audience interpret the mean effect size, we can present:

(1) Explanation for any significant moderators revealed by analyses.
(2) Describe the performance of any models you built in attempts to predict effect sizes.
(3) Discuss the diversity of the studies in sample.
(4) Consider the implications of findings for the major theoretical perspectives in the area of analysis.
(5) Make theoretical inferences based on results.
(6) Mention any features of analysis that might limit the generalizability of the results.
(7) Conclude with specific recommendations for the direction of future research.

15.3.4 REFERENCE AND APPENDIX SECTION

Have a single reference section that includes both studies used in writing the paper and those included in the meta-analysis. You should place an asterisk next to those studies included in the analysis.

You should prepare an appendix including all of the codes and effect sizes obtained in the analysis. Many journals will not be interested in publishing this information, but you will likely receive requests for it from people who read your report.

KEYWORDS

- reliability
- evidence
- applicability
- variations
- method section
- results section

- **discussion and conclusion section**
- **reference and appendix section**

REFERENCE

DeCoster, J. *Meta-Analysis Notes*, 2009. Retrieved on <August, 15, 2015> from http:// www.stat-help.com/notes.html.

CHAPTER 16

IMPLICATIONS OF RESULTS OF META-ANALYSIS

CONTENTS

ABSTRACT

Meta-analysis of existing evidence is prerequisite for scientific and ethical design of new studies. Research synthesis will not eliminate dissent or error, nor discourage innovation. The creation and wide spread use of synthetic tools in science will facilitate greater convergence in scientific communities. Meta-analysis helps clinicians to produce evidence based treatment. Applying results of meta-analysis to an individual patient involves, (1) consideration of the applicability of evidence, (2) the feasibility of the intervention in a particular settings, (3) risk ratio in the individual patient, and (4) incorporation of the patient values and preferences. The results can be extrapolated to a specific patient by either considering results in the most relevant subgroups or by multivariate risk prediction equations or by using clinical judgment to determine a specific patient's risk status. Meta-analysis is the optimum method of summarizing evidence of effectiveness of clinical practice guidance. Meta-analysis helps in the development of methods for economic evaluation. The results of meta-analysis helps in the determination of the number needed to treat.

Meta-analysis helps policy makers to make meaningful decisions. Policy making should be based on best current knowledge and takes into account of resources and values in the interpretation of evidence.

16.1 IMPLICATIONS FOR SCIENTISTS AND RESEARCHERS

16.1.1 PREREQUISITES FOR NEW STUDIES

Meta-analysis of existing evidence is prerequisite for scientific and ethical design of new studies. Proposals for new study should take into account of information about planned and ongoing studies. Ethical and scientific monitoring of ongoing studies should take account of systematic reviews that have incorporated relevant new evidence. The results of new studies should be set and interpreted in the context of systematic reviews of all of the relevant evidence available at the time of reporting. Up-to-date meta-analysis of existing evidence and registers of planned and ongoing studies are essential for scientific and ethical study design, monitoring and reporting, and for protecting the interests of consumers of research results (Egger et al., 2001).

16.1.2 ACCUMULATION OF KNOWLEDGE

Research synthesis will not eliminate dissent or error, nor discourage innovation. It cannot displace primary scientific inquiry. But it can uniquely reinvigorate the status of science as an objective, consensual enterprise that accumulates knowledge, whose end product is not just technology but more informed policy. How can we expect scientists even within narrow fields to come to wide consensus on anything, if there are no adequate or reliable vehicles for presenting actual evidence? And if consensus is not evident in the scientific community, how can we expect policy-makers or the public to agree on the implications of scientific research?

16.1.3 GREATER CONVERGENCE

The creation and widespread use of synthetic tools in science will facilitate greater convergence in scientific communities. As a corollary, the synthetic turn that the map of science has the potential to usher in will propel science into a new era, by shortening time-lag between paradigms, through reductions in the time that advocates of a position hold on to it.

16.2 IMPLICATIONS FOR CLINICIANS

Meta-analysis helps clinicians to produce evidence-based treatment. Applying results of meta-analysis to an individual patient involves (1) consideration of the applicability of evidence, (2) the feasibility of the intervention in a particular settings, (3) risk ratio in the individual patient, and (4) incorporation of the patient values and preferences. Even though meta-analysis provides the best estimate of the true effect of an intervention, their application at bedside is a difficult, time consuming, and incompletely studied skill (Egger et al., 2001). Within the process of guideline development, conducting specific meta-analysis or updating existing ones will usually represent the optimum means of summarizing evidence on the effects of interventions. Such reviews will be focused on the subject area of the guidelines and the need of the guideline group and can be tailored to the clinical questions that they pose.

16.2.1   APPLICABILITY

While some variation in treatment response between the patient and patients in the meta-analysis is to be expected, the differences tend to be quantitative rather than qualitative. The underlying pathobiology of the patient must be as that of the results of meta-analysis.

16.2.2   FEASIBILITY

This involves considerations of whether intervention is available and affordable in the settings and whether the necessary expertise and sources are totally available. The results of the local circumstances must also be considered. Again consider whether the necessary monetary facilities are available if the intervention is offered.

16.2.3   BENEFITS

In order to facilitate the extrapolation to a specific patient formats which incorporate base line risk and therapeutic effects are preferable.

The results can be extrapolated to a specific patient by either considering results in the most relevant subgroups or by multivariate risk prediction equations or by using clinical judgment to determine a specific patient's risk status.

Deriving clinically useful estimates of the overall results and extrapolating from the overall results to derive estimates for the individual patient.

The number needed to treat (NNT) and number needed to harm (NNH) are most important indices here: One can use multivariate risk prediction equations to quantitative individual patient's potential for benefit from the therapy. The clinician can divide the average NNT by a factor which relates the risk of individual patient to that of the average patient in the published report. If NNT is available, in various subgroups, the clinician can extrapolate the most relevant one for his patient.

16.2.4   INCORPORATION OF PATIENT'S VALUES

The techniques include patient's decision supporting technology or the expression of likelihood to help or harm.

16.2.5   CLINICAL GUIDELINES

Meta-analysis is the optimum method of summarizing evidence of effectiveness of clinical practice guidance. Within the process of developing a guideline, conducting specific systematic reviews or updating existing ones allows reviews to be focused on the subject area of the guideline and to be tailored to the clinical questions that group process.

There will be occasions when previously available meta-analysis will present the best available evidence. However, there may be problems in the interpretation and applicability of available meta-analysis.

16.3   IMPLICATIONS FOR ECONOMISTS

Meta-analysis helps in the development of methods for economic evaluation. The results of meta-analysis help in the determination of the NNT. The NNT is the number of patients that must be treated over a defined period of time to prevent one death/disorder. Hence, it is the reciprocal of the risk difference. Economic evaluation seeks to predict the net change in benefits and costs arising from alternative approaches to provide a particular form of care. Methods for economic evaluation and the results of meta-analysis can enrich each other. To improve information for optimal decisions in health care, results of meta-analysis of effectiveness should be used in economic evaluation.

16.4   NUMBER NEEDED TO TREAT

The NNT are commonly used to summaries the beneficial effects of the treatment in a clinically relevant way, taking into account the baseline risk without treatment and risk reduction achieved with treatment.

16.4.1   FACTORS

The NNT is sensitive to factors that change baseline risk, the outcome considered, characteristics of patient included in the trials, secular trend in the incidence and case fatality and finally the clinical settings.

16.4.2 RELEVANCE

Meaningful NNT is obtained by applying the pooled relative risk reduction calculated from meta-analysis or individual trials to the baseline risk relevant to specific patient groups.

16.4.3 EASY TO UNDERSTAND

They help to translate trial effects to clinical practice in terms that the clinicians can understand. High-risk patients to gain more from treatment and this is reflected in a small NNT, whereas low risk patients will have a large NNT. As absolute levels of risk are taken into account, the clinician can better weigh up the sizes of benefits with possible harms of treatment. Hence, NNT sets priorities. The NNT gives some idea of the clinical workload required to achieve health benefits and consequently valued by public health medicine as investment often seems disproportionate to the benefits obtained. The NNT is thought to be more intuitive and easier for clinicians to understand than relative measure of treatment effects.

16.4.4 LIMITATIONS

It may misleading because variation in event rates in trials, difference in the outcomes considered, effects of geographic and secular trends and this is risk, and the clinical settings. It is assumed throughout that, relative measures of treatment effects such as the odds ratio, risk ratio are the most appropriate measure of meta-analysis of trials. The NNT should be derived by applying relative risk reductions of treatment estimated by trials are meta-analysis to populations of specified absolute high, medium, or low risk to illustrate a range of possible NNT.

Mathematical aspect: If the treatment appears to have no effect, that is, the event rates are identical in both the treatments and control group, then the absolute risk difference is zero. In this case, the reciprocal of zero is infinite. It is rather confusing to have a measure of effect with confidence intervals that may include benefits and infinite.

Clinical settings: The relative estimate of efficacy varied less across the different settings and could be generalized with more confidence.

Assumptions: Decisions affecting the baseline risk of patients in a trial such as inclusion and exclusion criteria or geographical settings are not made in a random way.

Standardization: Trials have different levels of follow-ups. In order to produce, for example, 1-year NNT, all the absolute risk differences need to be standardized for 1 year if pooling of risk difference is undertaken. This standardization requires an assumption of constancy of effect over time.

Interpretation: In the economic field, an incremental cost-effectiveness analysis of an intervention at different levels of baseline risk will almost always be more informative than a summary of cost effectiveness based on a pooled NNT. The pooled NNT may also results in erroneous decisions about who should receive treatment in concept of a threshold NNT, separately those who are likely to benefit from those who are not, is applied.

Cohort studies: It is preferable to derive NNT from prognosis: from cohort studies rather than from the trials and meta-analysis themselves.

16.5 IMPLICATIONS FOR POLICY MAKERS

Meta-analysis helps policy makers to make meaningful decisions. Policy making should be based on best current knowledge and takes into account of resources and values in the interpretation of evidence. Policy making, like all other health-care decisions should be based on best current knowledge. Interpretation of evidence by policy makers also take into account resources and values. The job of the scientists is to clear about the evidence. Having received the evidence, decision will be taken by those who represent the values of society, which has to be respected by scientists irrespective of outcome (Egger et al., 2001).

The policy makers have to take a decision as to how much gross national product should be invested in public services, how much of the money should be spent on improvement of health according to geographical and different categories of health.

In some decisions where resources are not a major issue and the values are relatively straight forward, policy decisions can be based on evidence alone.

KEYWORDS

- **implications**
- **scientists**
- **clinicians**
- **policy makers**
- **NNT**

REFERENCE

Egger, M.; Smith, G. D.; Altman, D. G., Eds. *Systematic Reviews in Health Care—Meta-analysis in Context*. BMJ Publishing Group: London, 2001.

CHAPTER 17

META-ANALYSIS SOFTWARE

CONTENTS

ABSTRACT

Computer software entirely devoted to meta-analysis has increasingly become available, and meta-analytic procedures have also been introduced in general statistical software packages. The public domain software is freely downloadable along with user manual. Majority of software under this category are DOS based and provide basic meta-analytical procedures. The portability of data as well as graphics produced is less flexible with this software. The public domain meta-analysis software is compared based on their utility, suitability, portability, ease in operation, graphical representation of results, and scope for upgradation. The commercial software has to be procured at appropriate cost. The main advantage of this software is that, most of them are Window based and allow easy transfer of data, results, and graphics from one platform to another.

17.1 PUBLIC DOMAIN SOFTWARE

Over the past few years, computer software entirely devoted to meta-analysis has increasingly become available, and meta-analytic procedures have been introduced in general statistical software packages (Kuss and Koch, 1996).

The public domain software is freely downloadable along with user manual. Majority of software under this category are DOS based and provide basic meta-analytical procedures. The portability of data as well as graphics produced is less flexible with this software. The examples for these are RevMan, EasyMa, and meta-analyst.

The comparative study of public domain software facilitates the researcher to identify the best software in general and medical research in particular. The public domain meta-analysis software is compared based on their utility, suitability for prevalence-based meta-analysis. The portability, ease in operation, graphical representation of results, and scope for upgradation will be considered.

17.1.1 REVMAN

The Cochrane Collaboration's Review Manager (RevMan) is a software package designed to enter review protocols or completed reviews in

Cochrane format. This includes a structured text of the review and tables of included as well as excluded studies. The program is based on the Windows operating system and is easy to use. Dichotomous or continuous data can be entered and analyzed in fixed and random-effects models for odds ratio, relative risk, risk difference, and weighted mean difference. Different comparisons and outcomes can be accommodated in the same data sheet. The classical meta-analysis graph is displayed with or without raw data, weights, and year of individual studies. The graphics can be edited on the screen and printed.

17.1.2 EASYMA

EasyMA was developed by Michel Cucherat from the University of Lyon and can be down loaded from an Internet. It is menu driven and offers fixed-effects and random-effects models. The number of patients needed to treat to prevent one event is also given. EasyMA produces the classical meta-analysis graphs both for standard and cumulative meta-analysis as well as radial and funnel plots. Rosenthal's number of unpublished negative trials needed to render the combined results nonsignificant and Begg and Mazumdar's test for publication bias are also available.

Joseph Lau from the New England Medical Center programmed meta-analyst. Like EasyMA, this software was developed for conventional and cumulative meta-analysis of clinical trials with two arms and dichotomous outcomes. Only one outcome can be entered at a time. The program is easy to use and offers the widely used fixed effects and random effects models. The graphs produced are of excellent quality, but as in EasyMA, they cannot easily be exported for editing in another program.

17.2 COMMERCIAL SOFTWARE

The commercial software has to be procured at appropriate cost. The main advantage of these software is that, most of them are Window based and allow easy transfer of data, results, and graphics from one platform to another. The example for these are FAST*PRO, STATA, TrueEpistat, DSTAT, and DESCARTES.

A few commercially available general statistical software like STATA, SAS, S-Plus, StatsDirect, StatXact, and TrueEpistat have included facilities for meta-analysis by introducing specialized routines (Kuss and Koch, 1996). The commercial software is bound to make frequent changes to appeal the customers with an additional cost. While choosing the best software, one has to compare the facility provided against cost (Egger et al., 1998).

17.2.1   FAST*PRO

The FAST*PRO is entirely devoted to the appraisal of evidence. Results from a wide range of experimental and observational study designs can be analyzed using dichotomous, categorical, or continuous effect measures, including odds ratios, relative risks, and risk differences. The program accommodates studies conducted in single groups of patients (e.g., natural history studies) and studies in which groups received different doses of the same treatment. The software is based on the confidence profile method, which uses Bayesian statistics to calculate posterior probability distributions for parameters of interest. The probability distributions, referred to as profiles, are graphically displayed to provide a visual picture of the uncertainty attached to a parameter. Different models can be used to combine the profiles constructed from individual studies. This includes Bayesian models that accommodate the fixed and random-effects assumption, conventional fixed-effects models (e.g., variance-weighted, Mantel–Haenszel, Yusuf–Peto) and the DerSimonian–Laird random effects model.

17.2.2   STATA

STATA is a comprehensive statistics, data management and graphics package for which, a meta-analysis command has recently been written. Individual-level or study-level data are analyzed using standard methods to provide an effect estimate (e.g., odds ratio, risk difference, or difference between means) and corresponding standard error for each study. The meta-analysis command then calculates fixed-effects (variance-weighted) and random-effects (DerSimonian and Laird, 1986) estimates, together with the standard test for heterogeneity between studies and estimate of between-study variance. The classical meta-analysis graph is displayed with either the fixed-effects or random-effects combined estimate. Empirical Bayes estimates of the true effect in each study given the random-effects model can be calculated, displayed, and graphed. Results and graphical displays can be shown either on the original scale or on the ratio scale (when the original effect estimates are on a log scale). Funnel plots can be displayed using the standard graphics facilities of the package. Commands on meta-regression, which can be used to explore sources of heterogeneity between studies, and on the cumulative meta-analysis are also available.

17.2.3   TRUE-EPISTAT

True-Epistat is a statistics package, which also offers a number of meta-analysis capabilities. Studies comparing two groups and using odds ratios, relative risk, risk difference (dichotomous outcomes), or standardized differences (continuous outcomes) can be analyzed in a variance-weighted fixed effects model or a DerSimonian–Laird random effects model. Data are entered in a two-by-two table or as group variances along with the difference between two means. Correlation coefficients, test statistics from widely used distributions and p-values (and mixtures of the former) can also be combined. The results are given in tabular form or in the typical graphical display showing effect measures and confidence intervals for each study and for the overall result. Funnel plots can be drawn. The graphics can be edited on the screen and printed.

17.2.4   DSTAT

DSTAT was developed for meta-analysis in the psychological sciences. It combines studies comparing two groups. The data are entered as correlation coefficients, test statistics, p-values, or mixtures. These statistics are then converted into a standardized (scale-free) effect measure, the effect size (Hedges g) that is defined as the difference between the two groups expressed in pooled standard deviation units. If the user wishes to do so, a bias-adjusted effect size can be calculated. Adjusted or unadjusted effect sizes of individual studies are then combined to produce an overall value. Clinically more relevant quantities such as the difference in risk, the relative risk, or the odds ratio cannot be calculated with DSTAT. Also, results cannot be graphically displayed. These drawbacks limit the usefulness of DSTAT for meta-analysis in medical research.

17.2.5   DESCARTES

DESCARTES is a set of software tools for writing systematic reviews and performing meta- analysis, which is being developed by Update Software, the company, which is responsible for the Cochrane Library and the Cochrane Collaboration's RevMan package. It is an interactive guide through all the steps involved, from protocol through data collection and analysis to publication in paper and electronic formats. The emphasis is on producing a finished document (protocol, systematic review, or individual meta-analysis). Output is geared toward publication of quality graphics, text, and tables. Data for meta-analyses can be imported from other packages (e.g., RevMan), entered directly into a spreadsheet or entered interactively via guided data-entry screens, which check for impossible or unlikely values. Standard fixed effects and random effects models and meta-regression models are available for dichotomous, continuous, and individual patient data. Graphical output allows a wide range of plots (for example Funnel, L'Abbe, and Galbraith plots). Cumulative meta-analysis and sensitivity analyses are also available. Suggested interpretations of calculated statistics can be generated automatically in textual form and included in the output. All DESCARTES output can be pasted directly into other Windows packages.

17.3   META-ANALYSIS CALCULATOR

In addition to the statistical software available, a meta-analysis calculator was developed using a Microsoft Excel by Hanji et al. (2006). In this calculator, two different methods were incorporated mainly, the IV method meta-analysis and DL method meta-analysis with incorporation of fixed effects model and random effects model, respectively. This calculator has been validated by comparison with the results obtained by the STATA. It is worthy of stating here that, one of the lacunae in the literature was that the researchers employed available packages rather than developing the suitable calculators. To develop this, calculator ideas and sometimes sub-routines were borrowed from various sources (Egger et al., 2001).

KEYWORDS

- **public domain software**
- **commercial software**
- **STATA**

REFERENCES

DerSimonian, R.; Laird, N. Meta-analysis in Clinical Trials. *Controlled Clin. Trials* **1986,** *7*, 177–188.

Egger, M.; Schneider, M.; Smith, G. D. Meta-analysis: Spurious Precision? Meta-analysis of Observational Studies. *Br. Med. J.* **1998,** *316*, 140–144.

Egger, M.; Smith, G. D.; Altman, D. G., Eds. *Systematic Reviews in Health Care—Meta analysis in Context.* BMJ Publishing Group: London, 2001.

Hanji, M. B.; Suresh, K. P.; Reddy, M. V. *Meta-analysis Calculator.* Department of Biostatistics, NIMHANS: Bangalore, 2006.

Kuss, O.; Koch, A. Meta-Analysis Macros for SAS. *Stat. Softw. Newslett.* **1996,** *22*, 325–333.

CHAPTER 18

RUNNING META-ANALYSIS USING STATA

CONTENTS

ABSTRACT

STATA is a general-purpose, command-line driven, programmable statistical package in which commands to perform several meta-analytic methods are available. The major advantage with STATA is the continuous updating of the user written commands and placing them online along with complete description of the same including its utility. They are freely available on the Internet. All the commands related to meta-analysis can be searched using the command "search meta." The editor can be accessed by issuing a dot command "edit" or by clicking data in the main menu and data editor in the sub menu. Later data can be entered or edited in any of the cell by assuming every column is one variable and every row is observation for that particular variable. The major commands like meta for point estimators, metan for ratio estimators, funnel for funnel plot, labbe for labbe plot and metacum for cumulative meta-analysis are illustrated using live examples, syntax of commands and their variations to get desired output along with proper interpretation. Similarly other commands like metap for mata-analysis of p values, metacum for cumulative pooled estimates, metareg for meta-regression, metainf for influence analysis, metabias for publication bias, galbr for Galbraith plot for heterogeneity and some more commands are used as per the syntax provided to complete the meta-analysis.

18.1   GETTING STARTED

STATA is a general-purpose, command-line driven, programmable statistical package in which commands to perform several meta-analytic methods are available. The major advantage with STATA is the continuous updating of the user written commands and placing them online along with complete description of the same including its utility. They are freely available on the internet (Egger et al., 2001).

Soon after installing the core STATA software at our computer, we need to download all the user written commands published in the STATA Technical Bulletin by executing a dot command, namely, "**update all.**" This being an online command, we have to ensure internet connectivity to our computer before issuing the above command. This command will automatically connect our system to www.stata.com and update the core package with latest developments on user written commands.

All the commands related to meta-analysis can be searched using the command "search meta." This will list out all meta-analysis-related user written commands along with the STATA technical bulleting serial number starting with stb. The specific stb commands can be installed by selecting help menu followed by clicking on "STB and User-Written Programs," "http://www.stata.com," "stb," and select a specific stb command listed and click to install the same.

To get any help on the specific stb command including its syntax, utility, etc., we can issue a dot command **"help"** followed by specific stab command's name in a line. This will educate us in full about any stb command.

18.2 PREPARATION OF DATA USING THE EDITOR

The editor can be accessed by issuing a dot command **"edit"** or by clicking Data in the main menu and Data editor in the sub menu. Later data can be entered or edited in any of the cell by assuming every column is one variable and every row is observation for that particular variable. By default variable names are named at var1, var2, var3, etc. By clicking on the default variable name given by the system, we can change the variable name as per our requirement. By selecting all the observations in a variable or a single observation under any variable, we can delete an entire variable or particular observation as per our requirement after choosing the icon delete provided just above the editor. We can enter the data given in the following table using the edit command.

Sl. No.	Study name	Study year	Effect size	Standard error	Variance
1	Reddy	2010	0.4	0.05	0.0025
2	Hanji	2012	0.8	0.30	0.0900
3	Patil	1945	0.5	0.20	0.0400
4	Gowda	1960	0.7	0.25	0.0625
5	Swamy	2001	0.6	0.15	0.0225
6	Singh	2005	0.9	0.40	0.1600

The diagrammatic view of the STATA package when we completed the data entry is given below.

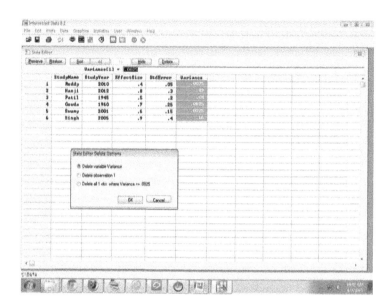

The date entered in the editor can be saved to the file with any name using the following command

save filename

The data will be saved to the filename specified in the above command with extension dta.

The stored file can be opened by using the command

use filename

We can close the editor by clicking on the close icon

18.3   RUNNING THE COMMAND META

The command "**meta**" expects the user to provide estimator along with either standard error or variance of the estimator for each study. It calculates pooled estimate under fixed effects/random effects/Bayesian approach using inverse-variance weighting method. It also tests the true pooled effect is zero or not andprovides confidence limits. A test for heterogeneity between studies is conducted after estimating the between studies variance and optionally, plots the individual and pooled estimates in forest plot.

As there is a separate command, namely, "metan" available exclusively for ratio estimators like odds ratio, etc., our discussion here is restricted to options of effect size only.

The syntax of the command is as following

meta Variable name containing estimator Variable name containing standard error of estimator OR Variable containing variance of estimator with following options:
[, var print ebayes level(#) graph(f|r|e) xlabel() id(strvar) fmult(#) boxysca(#) boxshad(#) cline ltrunc(#) rtrunc(#)]

The details of options are explained below:

, **var** Soon after second variable followed by ", var" means second variable is not a standard error of the estimator but that contains variance of the estimator.

print produces weights and confidence intervals for individual studies.

ebayes level(#) By default # has to be 95 and it creates two variables, namely, ebest and ebse and ebest contains Bayes estimate and ebse contains standard error for individual study.

graph(f|r|e) produces forest plot with combined estimate based on the option selected f,r,e for fixed effect, random effect, Bayesian estimate, respectively.

xlabel(values separated by cama) Provides labels for X-axis.

id(strvar) Here, server is the name of the variable containing name of the individual study.

fmult(#) decides the font size for Study labels of graphs by fixing # with any number (0–8).

boxysca(#) decides vertical length of the graph by fixing # as 1 or 0.

boxshad(#) decides shade of box of graph by fixing # as 0–4.

cline prints vertical dotted line at the pooled estimate value.

ltrunc(#) truncates the left side of the graph at the value specified at # which must be less than all values of effects sizes of individual studies.

rtrunc(#) truncates the right side of the graph at the value specified at # which must be greater than all values of effects sizes of individual studies.

Now, we can try above command with all possible combinations with the above data and view the results one by one.

1. meta EffectSize StdError

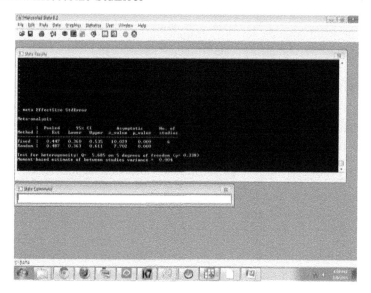

2. meta EffectSize Variance, var

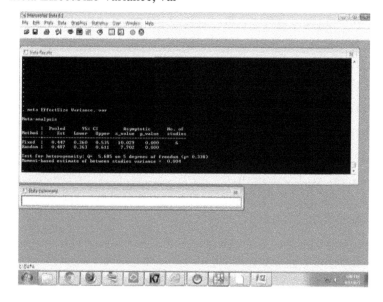

3. meta EffectSize Variance, var print

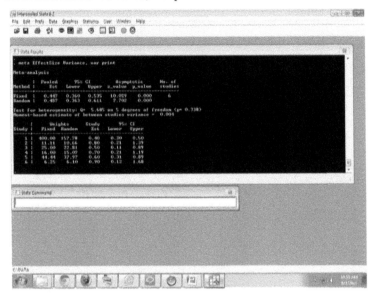

4. meta EffectSize Variance, var ebayes print

The editor containing newly created variables, namely, ebest and ebse is displayed below

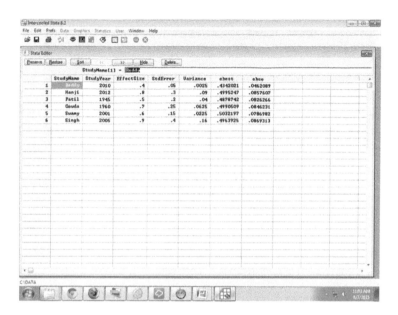

5. meta EffectSize Variance, var graph(f)

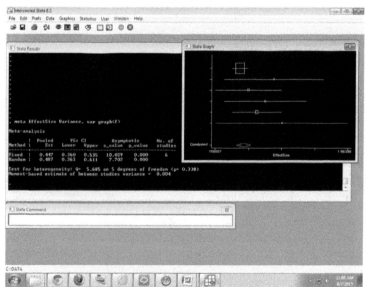

6. meta EffectSize Variance, var graph(r)

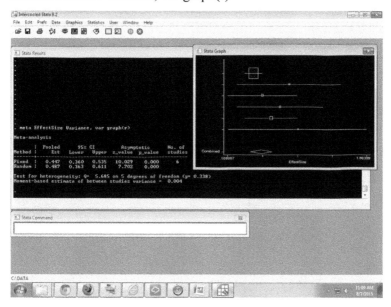

7. meta EffectSize Variance, var graph(e)

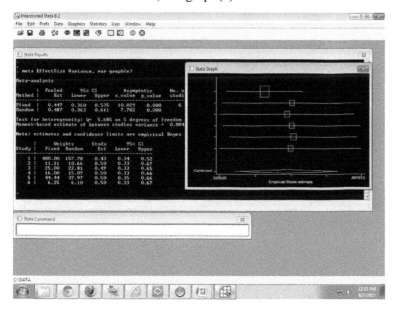

8.   meta EffectSize Variance, var graph(e) xlabel(.3,.4,.5,.6,.7)

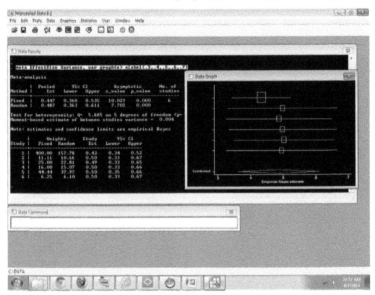

9.   meta EffectSize Variance, var graph(e) xlabel(.3,.4,.5,.6,.7) id
     (StudyName)

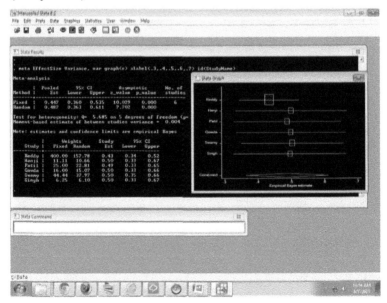

10. meta EffectSize Variance, var graph(e) xlabel(.3,.4,.5,.6,.7) id(Study Name) fmult(2)

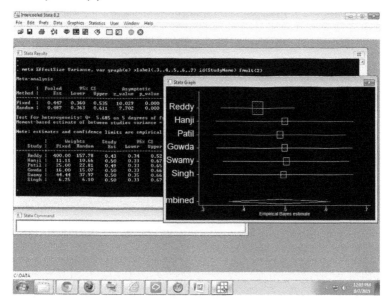

11. meta EffectSize Variance, var graph(e) id(StudyName) boxysca(.3)

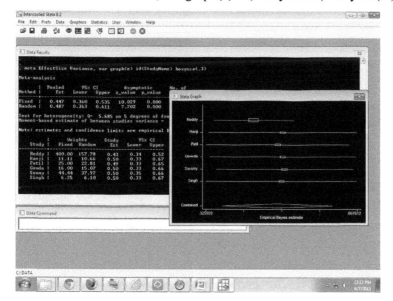

12. meta EffectSize Variance, var graph(e) id(StudyName) boxshad(4)

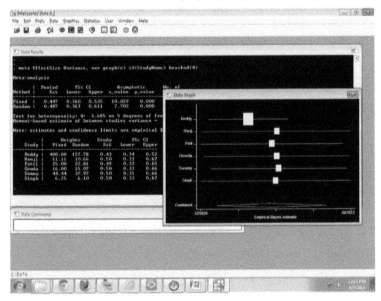

13. meta EffectSize Variance, var graph(e) id(StudyName) boxshad(4) cline

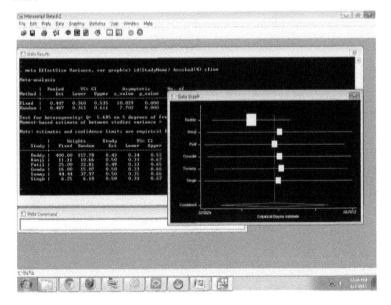

14. meta EffectSize Variance, var graph(e) id(StudyName) boxshad(4) cline ltrunc(.4)

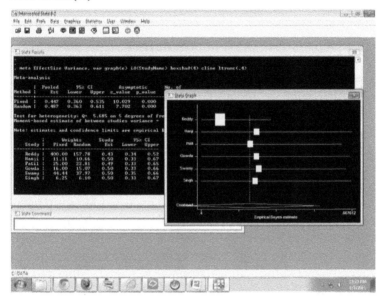

15. meta EffectSize Variance, var graph(f) id(StudyName) boxshad(4) cline rtrunc(.9)

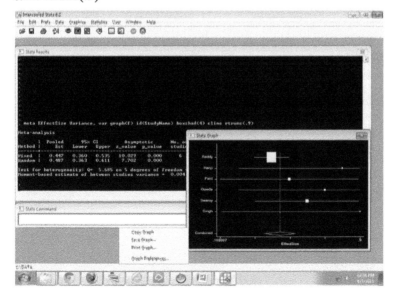

We can save or copy or print the graph by clicking right switch of the mouse as shown above.

18.4   RUNNING THE COMMAND METAN

The command "metan" expects user to provide either four or six variables. When four variables are provided, ratio estimators like odds, risk, or risk difference estimators are calculated. Six variables means data is expected to come from two groups with estimator, its standard error and sample size for each group. The command calculates pooled estimate under fixed effects/random effects/Bayesian using inverse-variance weighting and provides confidence limits and tests the true pooled effect is zero. A test for heterogeneity between studies is conducted after estimating the between studies variance and optionally, plots the results in forest optionally to describe heterogeneity. In addition, L'Abbe plot and funnel plots can be produced to investigate heterogeneity and bias among the studies as per requirement with additional commands, namely, labbe and funnel.

The syntax of the command is as following:

metan names of variables [, rr or rd hedges cohen glass nostandard fixed fixedi random randomi peto chi2 ilevel(#) olevel(#) notable nograph xlabel(#,...,#) force boxsha(#) boxsca(#) texts(#) saving(filename) nowt nostats nooverall].

funnel Names of variables [, sample noinvert ysqrt overall(#) graph options].

labbe Names of variables [, nowt percent graph options].

The details of options are explained below:

Options for binary data

rr pools risk ratios.

or pools odds ratios.

rd pools risk differences.

fixed specifies a fixed effect model using the Mantel–Haenszel **method.**

fixedi specifies a fixed effect model using the inverse variance method.

random specifies a random effects model using the method of DerSimonian & Laird, with the estimate of heterogeneity being taken from the Mantel–Haenszel method.

randomi specifies a random effects model using the method of DerSimonian & Laird, with the estimate of heterogeneity being taken from the inverse variance method.

peto specifies that Peto's assumption free method is used to pool odds ratios.

chi2 displays the chi-squared statistic (instead of z) for the test of significance of the pooled effect size. This is available only for odds ratios pooled using Peto or Mantel–Haenszel methods.

Options for continuous data

cohen pools standardized mean differences by the method of Cohen.

hedges pools standardized mean differences by the method of Hedges.

glass pools standardized mean differences by the method of Glass

nostandard pools unstandardized mean differences.

fixed specifies a fixed effect model using the inverse variance method with Cohen mean difference by default.

random specifies a random effects model using the method of DerSimonian & Laird with Cohen mean difference by default.

General output options

ilevel() specifies the significance level for the individual trial confidence intervals, namely, 95, 99.

olevel() specifies the significance level for the overall (pooled) confidence intervals, namely, 95, 99.

notable prevents display of table of results.

nograph prevents display of graph.

Graphical display options for forest plot

boxsha() controls box shading intensity, between 0 and 4.

boxsca() controls box scaling, which by default is 1.

texts() specifies font size for text display on graph. The default size is 1.

nowt prevents display of study weight on graph.

nostats prevents display of study statistics on graph.

nooverall prevents display of overall effect size on graph: default enforces the nowt option

Options for funnel

The funnel command with no parameters specified will produce a standard funnel plot of precision (1/SE) against treatment effect.

sample denotes that the y-axis is the sample size and not a standard error.

noinvert prevents the values of the precision variable from being inverted.

ysqrt represent y-axis on square root scale.

overall(x) draw a dashed vertical line at overall effect size given by x.

Options for labbe

By default the size of the plotting symbol is proportional to the sample size of the study. If weight is specified the plotting size will be proportional to the weight variable.

nowt declares that the plotted data points are to be the same size.

percent display the event rates as percentages rather than proportions.

xlabel(values separated by cama) display the values on x-axis.

ylabel(values separated by cama) display the values on y–axis.

We will enter the following binary data consisting of four cells for each study depicting effect or no effect in intervention and control groups into the editor before executing the metan command.

Sl. No.	Study name	Study year	Intervention group		Control group	
			Effect	No effect	Effect	No effect
1	Reddy	2010	1	39	2	34
2	Hanji	2012	9	126	23	112
3	Patil	1945	2	198	7	193
4	Gowda	1960	1	47	1	45
5	Swamy	2001	10	140	8	140
6	Singh	2005	1	58	9	47

After entering the data with suitable variable names, the STATA editor looks like below.

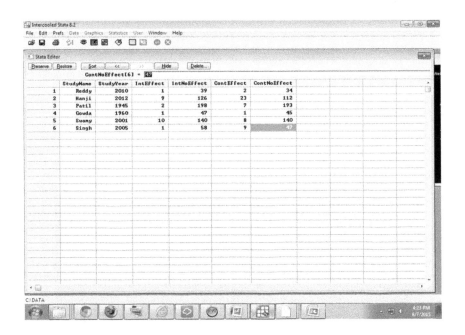

Now, we will execute all possible combinations metan command for binary data given above and view the results:

1. metan IntEffect IntNoEffect ContEffect ContNoEffect, rr

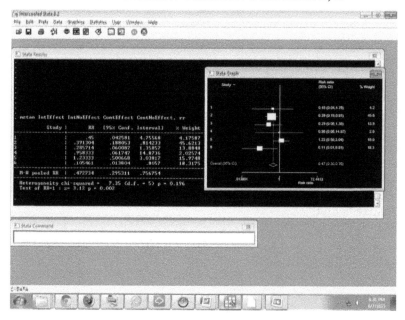

2. metan IntEffect IntNoEffect ContEffect ContNoEffect, or

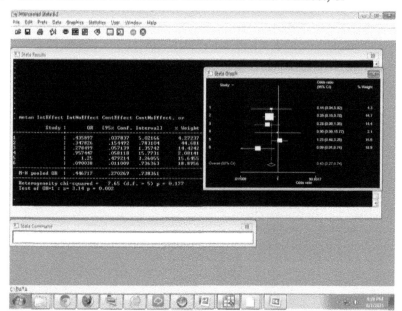

3. metan IntEffect IntNoEffect ContEffect ContNoEffect, rd

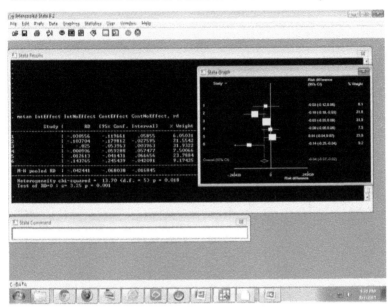

4. metan IntEffect IntNoEffect ContEffect ContNoEffect, or fixed

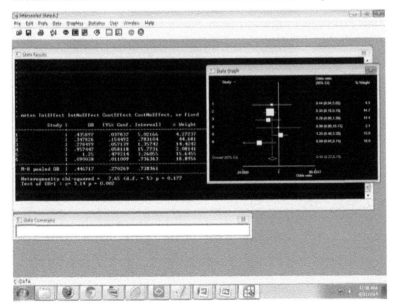

5. metan IntEffect IntNoEffect ContEffect ContNoEffect, or fixedi

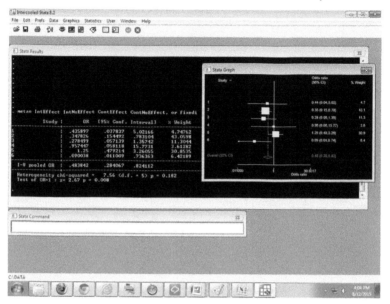

6. metan IntEffect IntNoEffect ContEffect ContNoEffect, or random

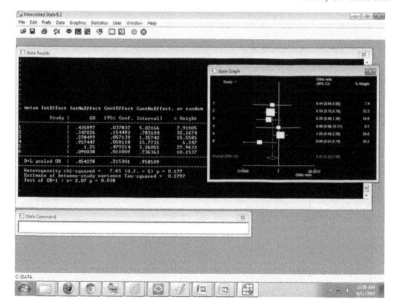

7. metan IntEffect IntNoEffect ContEffect ContNoEffect, or randomi

8. metan IntEffect IntNoEffect ContEffect ContNoEffect, or peto

9.   metan IntEffect IntNoEffect ContEffect ContNoEffect, or chi2

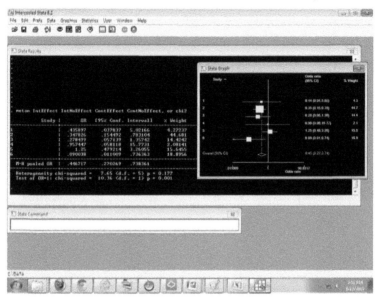

Now, we will enter the following continuous data consisting of six cells for each study depicting sample size, mean, standard deviation of the mean for both intervention and control groups into the editor before executing the metan command.

Sl. No.	Study name	Study year	Intervention group			Control group		
			Sample size	Mean	Standard deviation	Sample size	Mean	Standard deviation
1	Reddy	2010	6	39	1.3	7	34	2.4
2	Hanji	2012	9	126	3.1	8	112	2.8
3	Patil	1945	4	198	1.5	5	193	1.8
4	Gowda	1960	5	47	1.9	4	45	1.4
5	Swamy	2001	8	140	2.9	9	140	3.0
6	Singh	2005	7	58	2.5	6	47	1.2

After entering the data with suitable variable names, the STATA editor looks like below.

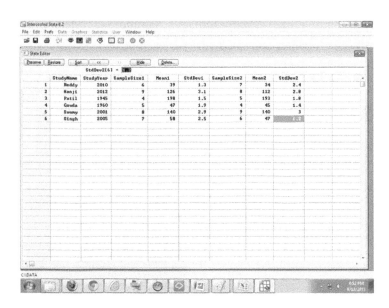

Now, we will execute all possible combinations metan command for continuous data given above and view the results:

10. metan SampleSize1 Mean1 StdDev1 SampleSize2 Mean2 StdDev2, cohen

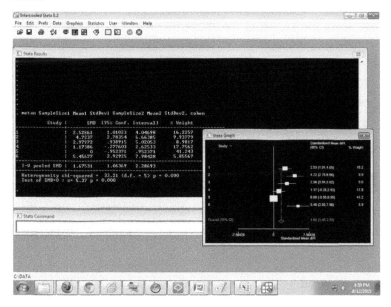

11. metan SampleSize1 Mean1 StdDev1 SampleSize2 Mean2 StdDev2, hedges

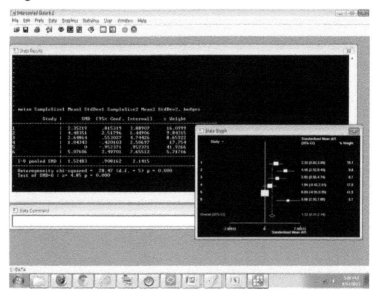

12. metan SampleSize1 Mean1 StdDev1 SampleSize2 Mean2 StdDev2, glass

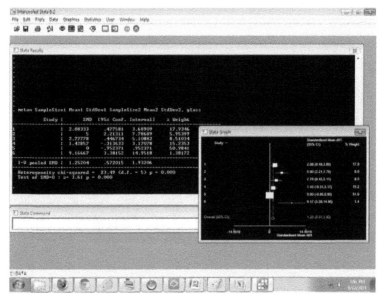

13. metan SampleSize1 Mean1 StdDev1 SampleSize2 Mean2 StdDev2, nostandard

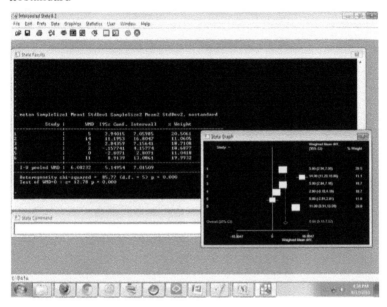

14. metan SampleSize1 Mean1 StdDev1 SampleSize2 Mean2 StdDev2, glass fixed

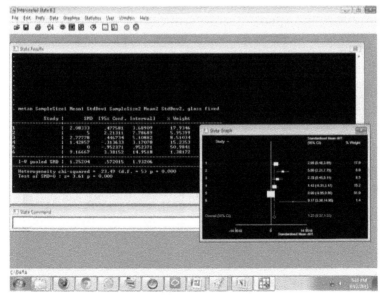

15. metan SampleSize1 Mean1 StdDev1 SampleSize2 Mean2 StdDev2, hedges random

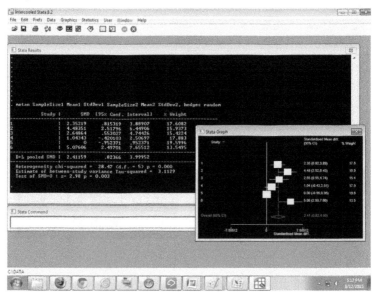

Now we will explore the general options available with metan command

16. metan SampleSize1 Mean1 StdDev1 SampleSize2 Mean2 StdDev2, glass-fixed ilevel(90)

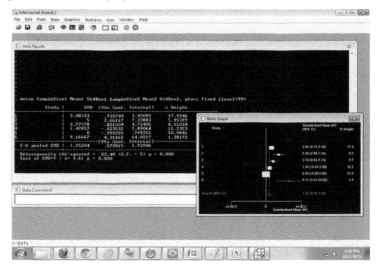

17. metan SampleSize1 Mean1 StdDev1 SampleSize2 Mean2 StdDev2, glass-fixed olevel(90)

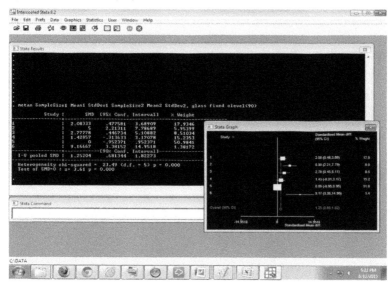

18. metan SampleSize1 Mean1 StdDev1 SampleSize2 Mean2 StdDev2, glass-fixed notable

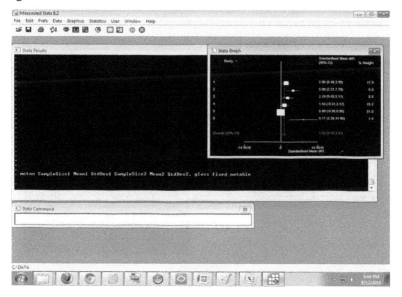

19. metan SampleSize1 Mean1 StdDev1 SampleSize2 Mean2 StdDev2, glass-fixed nograph

Now we will discuss the graph options available with metan command

20. metan SampleSize1 Mean1 StdDev1 SampleSize2 Mean2 StdDev2, glass-fixed boxsha(3)

21. metan SampleSize1 Mean1 StdDev1 SampleSize2 Mean2 StdDev2, glass-fixed boxsca(.5)

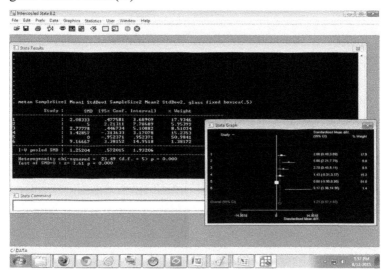

22. metan SampleSize1 Mean1 StdDev1 SampleSize2 Mean2 StdDev2, glass-fixed texts(.5)

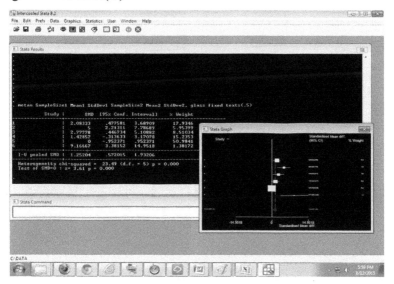

23. metan SampleSize1 Mean1 StdDev1 SampleSize2 Mean2 StdDev2, glass-fixed nowt

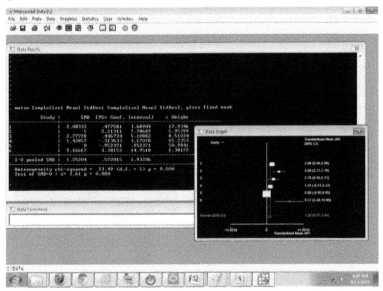

24. metan SampleSize1 Mean1 StdDev1 SampleSize2 Mean2 StdDev2, glass-fixed nostats

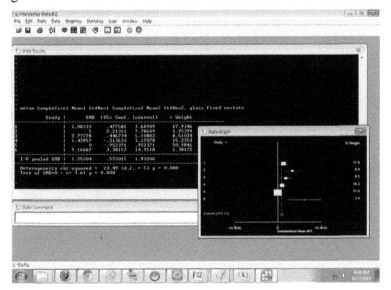

25. metan SampleSize1 Mean1 StdDev1 SampleSize2 Mean2 StdDev2, glass-fixed nooverall

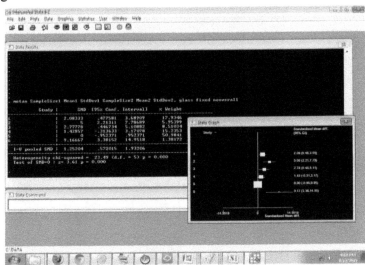

Now we will discuss the funnel graph options available with metan command

26. funnel,

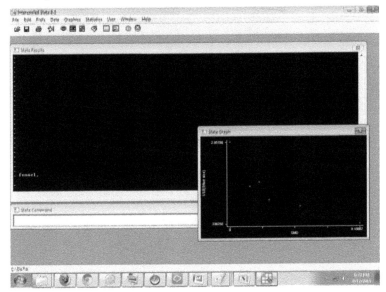

27. funnel, sample

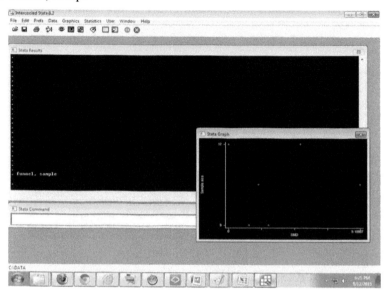

28. funnel, noinvert

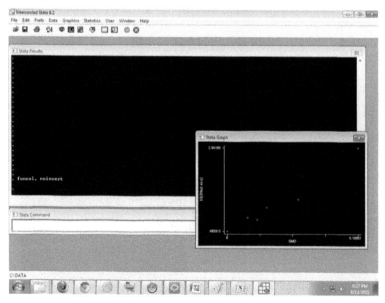

29. funnel, ysqrt

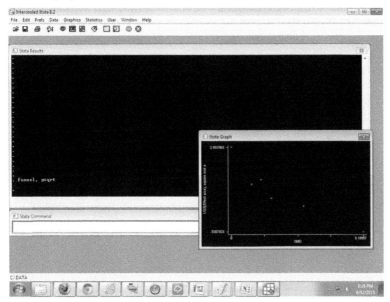

30. funnel, overall(1.25)

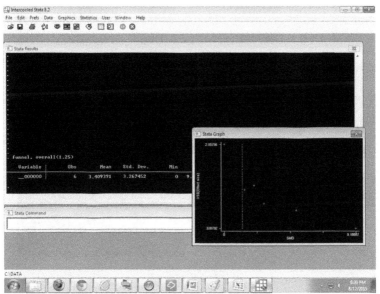

Now we will discuss the labbe graph options available with metan command but it requires the binary data only so we will activate our earlier data set as following

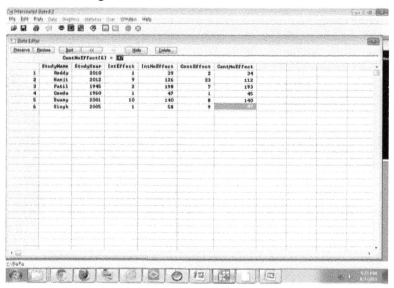

31. labbe IntEffect IntNoEffect ContEffect ContNoEffect,

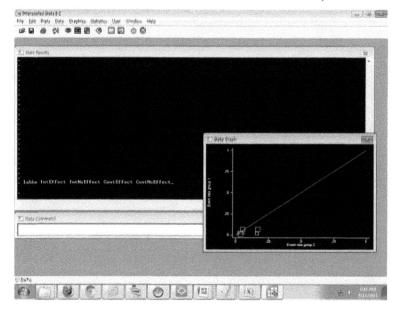

32. labbe IntEffect IntNoEffect ContEffect ContNoEffect, nowt

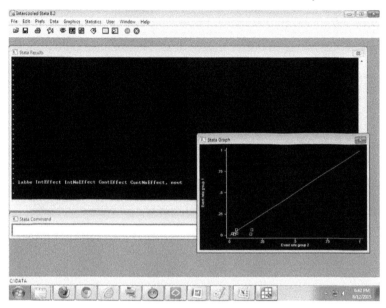

33. labbe IntEffect IntNoEffect ContEffect ContNoEffect, percent

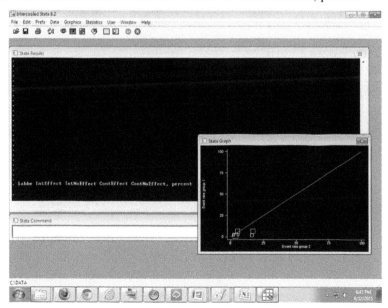

34. labbe IntEffect IntNoEffect ContEffect ContNoEffect, xlabel
    (0,.2,.4,.6,.8,1)

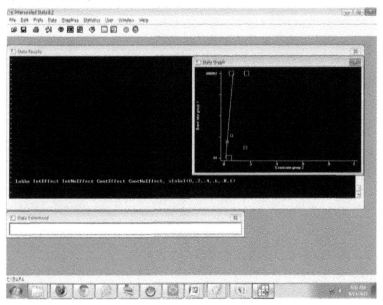

35. labbe IntEffect IntNoEffect ContEffect ContNoEffect, ylabel
    (0,.2,.4,.6,.8,1)

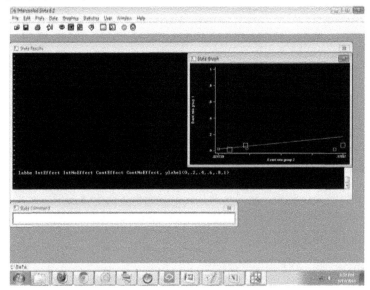

18.5   RUNNING THE COMMAND METACUM

The command "metacum" provides cumulative pooled estimates, confidence limits, and tests the true pooled effect is zero, obtained from fixed or random effects meta-analysis also, optionally, plots the cumulative pooled estimates. Command expects the user to provide estimator along with either standard error or variance of the estimator or lower and upper confidence intervals with confidence level (optional).

The syntax of the command is as following:

metacum Variable name containing estimator Variable name containing standard error of estimator OR Variable containing variance of estimator OR Lower confidence limit Upper confidence limit Confidence level(Optional), **var ci effect(f|r) eform graph id(strvar) cline csize(#)**

The details of options are explained below:

, var Soon after second variable followed by ", var" means second variable is not a standarad error of the estimator but that contains variance of the estimator

ci expects lower and upper confidence limits of the ratio scale estimator

effect(f|r) specifies provides fixed(f) or random-effects(r) estimates in the graph.

eform produces exponentiated estimates for ratio scale estimator in the log form for x-axis.

id(strvar) Here strver is the name of the variable containing name of the individual study

graph produces cumulative graph with regular graph options listed below

cline Prints vertical dotted line at the pooled estimate value

csize(#) with # specifies the size of the circles used in the graph (default 180)

Now we can try above command with all possible combinations with the below given data and view the results one by one.

Sl. No.	Study name	Study year	Effect size	Standard error	Variance	Odds ratio		
						Estimator	Lower CL	Higher CL
1	Patil	1945	0.5	0.20	0.0400	2.98	0.94	5.02
2	Gowda	1960	0.7	0.25	0.0625	1.17	0.28	2.63
3	Singh	2005	0.9	0.40	0.1600	5.46	2.93	7.89
4	Reddy	2010	0.4	0.05	0.0025	2.53	1.01	4.05
5	Hanji	2012	0.8	0.30	0.0900	4.72	2.78	6.66

After entering the data editor looks like below:

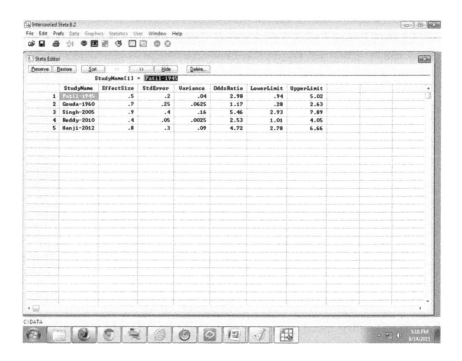

Now, we will execute all possible combinations metan command for continuous data given above and view the results

1. metacum EffectSize StdError, effect(f) graph

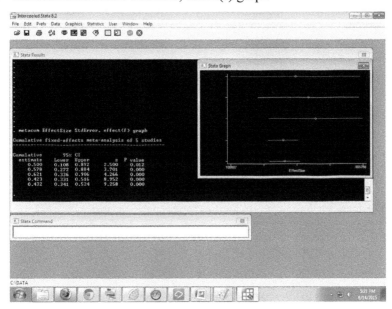

2. metacum EffectSize Variance, var effect(f) graph id(StudyName)

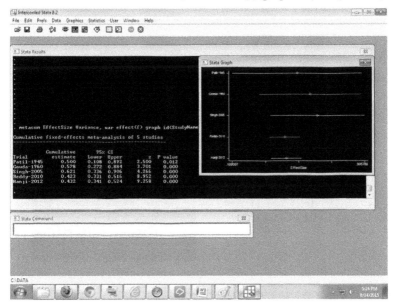

3. metacum EffectSize Variance, var effect(f) graph id(StudyName)
   cline

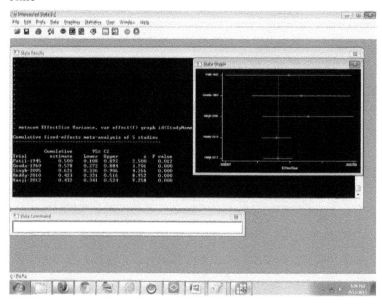

4. metacum EffectSize Variance, var effect(f) graph id(StudyName)
   cline csize(150)

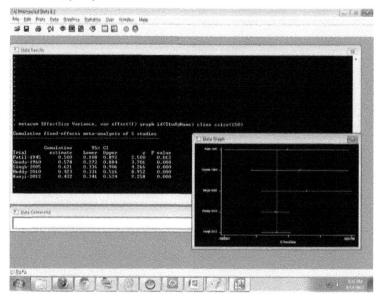

5. metacum OddsRatio LowerLimit UpperLimit, ci effect(f) graph id (StudyName) cline

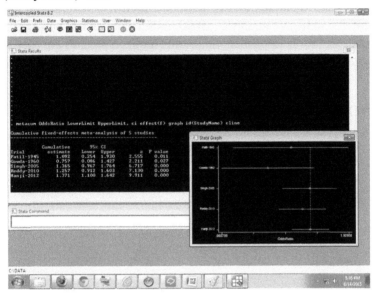

6. metacum OddsRatio LowerLimit UpperLimit, ci effect(r) graph id(StudyName) cline eform

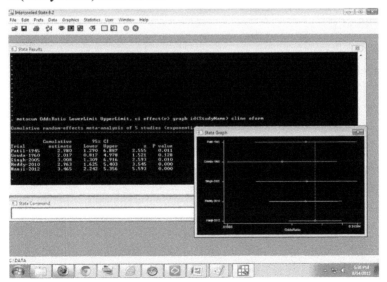

The above presentation is to have hands on experience of important commands of STATA for meta-analysis. Similarly other commands like **metap** for meta-analysis of p values, **metareg** for meta-regression, **metainf** for influence analysis, **metabias** for publication bias, **galbr** for Galbraith plot for heterogeneity and some more commands are used as per the syntax provided to complete the meta-analysis as per our requirement.

KEYWORDS

- meta
- metan
- funnel
- labbe
- metacum
- metareg
- metainf
- metabias

REFERENCE

Egger, M.; Smith, G. D.; Altman, D. G., Eds. Systematic Reviews in Health Care—Meta-analysis in Context. BMJ Publishing Group: London, 2001.

NUMERICAL DEMONSTRATION: META-ANALYTICAL APPROACH TO ESTIMATE PREVALENCE OF SCHIZOPHRENIA IN INDIA

CONTENTS

A.1   AIM AND OBJECTIVES

The study has two major objectives. First, it was aimed to conduct a meta-analysis to determine precise estimate of prevalence and pattern of schizophrenia in India. Second, to identify the suitable method among several approaches of meta-analysis under such situations based on sound statistical properties.

A.2   LOCATED STUDIES

The literature search for prevalence studies of schizophrenia have been started with search of Internet-based internationally acclaimed database MEDLINE through PubMed. There are 97 studies with the keywords "Prevalence, India, Schizophrenia." Some of the studies of Indian origin may not be indexed in MEDLINE, the database of Indian origin INDMED has been searched with the same keywords. There were only five studies with the keywords "Prevalence, India, Schizophrenia."

In addition to the above, more than 200 experts were identified in the field of schizophrenia, and e-mail was sent to them with a request to send unpublished, ongoing and MD and PhD thesis results to us. We were able to collect three MD thesis. We have also collected data from ICMR sponsored study reports. We have traced five ICMR sponsored studies for schizophrenia. The experts in the field of psychiatry have been requested to send the prevalence studies, which were unpublished, ongoing, MD and PhD dissertations through e-mail. The references of the leading articles in the prevalence of schizophrenia have been searched to locate the previous studies. This resulted in additional 33 studies for schizophrenia are there in Table A1.

TABLE A1   Source and Number of Schizophrenia Studies Located.

Source of location	Number of studies
MEDLINE	97
INDMED	5
Data from experts by MD thesis	3
Data from ICMR sponsored study reports	5
References of review articles and hand search of leading journals	33
Total number of studies located	143

A.3   CRITERIA FOR INCLUSION AND EXCLUSION OF STUDIES

Keeping the proposed objectives in mind, the following inclusion and exclusion criteria have been fixed to increase the homogeneity of studies and to make studies comparable. The inclusion criteria are as follows: (1) the core design of the study must be door-to-door enquiry, (2) availability of separate prevalence rate for schizophrenia, and (3) random selection of sampling units. The exclusion criteria are as follows: (1) hospital-based studies, (2) studies not covering all age groups, and (3) studies not reporting demographic characteristics of the population.

A.4   SELECTED SCHIZOPHRENIA STUDIES

The final selection of the studies according to the inclusion and exclusion criteria must be based on the assessment by two researchers blinding on the author, publication, and year of publication. If there were more than one publication on the same study, the study providing more information would be selected. There are 37 studies selected for meta-analysis as there in Table A2.

TABLE A2   List of Located/Selected Schizophrenia Studies for Meta-analysis.

Sl. No.	Chief investigator	Year of report	Name of journal/Source	Source of location	Remarks
1	Agarwal RB	1978	Gujarat University	MD Thesis	Selected
2	Agrawal P	1990	Acta Psychiatr Scand	MEDLINE	
3	Ali G	1997	Trop Med Int Health	MEDLINE	
4	Bagadia VN	1974	Int J Soc Psychiatry	MEDLINE	
5	Barry H	1967	Arch Gen Psychiatry	MEDLINE	
6	Batra L	1995	Indian J Psychiatry	INDMED	
7	Bhatia MS	2000	Indian J Med Sci	MEDLINE	
8	Bhatia MS	2000	Int J Soc Psychiatry	MEDLINE	
9	Bhatia SC	1987	Int J Soc Psychiatry	MEDLINE	
10	Bhatia T	2004	J Postgrad Med	MEDLINE	
11	Bhatia T	2004	Schizophr Res	MEDLINE	
12	Bhide A	1982	NIMHANS, Bangalore	MD Thesis	Selected
13	Bhugra D	1996	J Nerv Ment Dis	MEDLINE	

TABLE A2 *(Continued)*

Sl. No.	Chief investigator	Year of report	Name of journal/Source	Source of location	Remarks
14	Bhugra D	1999	Soc Psychiatry Psychiatr Epidemiol	MEDLINE	
15	Bolton P	1984	Int J Soc Psychiatry	MEDLINE	
16	Brown AS	1990	J Nerv Ment Dis	MEDLINE	
17	Campion J	1997	Soc Psychiatry Psychiatr Epidemiol	MEDLINE	
18	Carstairs GM	1976	The Universe of Kota (Book)	H S	Selected
19	Chakraborty A	1978	Ment Health Soc	MEDLINE	
20	Chandra PS	2003	Int J STD AIDS	MEDLINE	
21	Chandra PS	2003	Psychiatry	MEDLINE	
22	Chatterjee S	2003	Br J Psychiatry	MEDLINE	
23	Cheetham RW	1981	S Afr Med J.	MEDLINE	
24	Chong SA	2001	J Clin Pharmacol	MEDLINE	
25	Chopra GS	1974	Arch Gen Psychiatry	MEDLINE	
26	Cochrane R	1989	Soc Psychiatry Psychiatr Epidemiol	MEDLINE	
27	Cochrane R	1987	Soc Psychiatry	MEDLINE	
28	Collins PY	1996	Psychiatr Q	MEDLINE	
29	Das SK	1999	Soc Psychiatry Psychiatr Epidemiol	MEDLINE	
30	Dean G	1981	Br J Psychiatry	MEDLINE	
31	Deshpande SN	2004	Soc Psychiatry Psychiatr Epidemiol	MEDLINE	
32	Dhavale HS	2004	J Psychiatr Pract	MEDLINE	
33	Doongaji DR	1982	J Clin Pharmacol	MEDLINE	
34	Dube KC	1984	Acta Psychiatr Scand	MEDLINE	
35	Dube KC	1972	J Biosoc Sci	MEDLINE	
36	Dube KC	1970	Acta Psychiatr Scand	MEDLINE	Selected
37	Eaton WW	1995	Arch Gen Psychiatry	MEDLINE	
38	Eaton WW	1998	J Nerv Ment Dis	MEDLINE	
39	Elnagar MN	1971	Br J Psychiatry	H S	Selected
40	Floru L	1975	Confin Psychiatr	MEDLINE	
41	Gangadhar BN	2002	Acta Psychiatr Scand	MEDLINE	

TABLE A2   *(Continued)*

Sl. No.	Chief investigator	Year of report	Name of journal/Source	Source of location	Remarks
42	Ganguli HC	1968a	Ind J Med Research	H S	
43	Ganguli HC	1968b	Ind J Med Research	H S	
44	Ganguli HC	1968c	Ind J Med Research	H S	
45	Ganguli HC	2000	Indian J Psychiatry	H S	
46	Giggs J	1973	Nurs Times	MEDLINE	
47	Gopinath PS	1968	Bangalore University	MD Thesis	Selected
48	Goswami S	2003	Indian J Med Sci	MEDLINE	
49	Gupta S	1992	Br J Psychiatry	MEDLINE	
50	Hale AS	1994	Br J Psychiatry	MEDLINE	
51	Hambrecht M	1992	Eur Arch Psychiatry Clin Neurosci	MEDLINE	
52	Hayati AN	2004	Med J Malaysia	MEDLINE	
53	Henderson S	1976	Med J Aust	MEDLINE	
54	Isaak MK (Ed.)	1990	ICMR CAR CMH Study, Bangalore, Karnataka	ICMR-CAR	Selected
55	Issac MK	1980	Br J Psychiatry	H S	Selected
56	Issac MK (Ed.)	1987a	ICMR-DST, New Delhi Report from Gujarat	ICMR-DST	Selected
57	Issac MK (Ed.)	1987a	ICMR-DST, New Delhi Report from Karnataka	ICMR-DST	Selected
58	Issac MK (Ed.)	1987a	ICMR-DST, New Delhi Report from Punjab	ICMR-DST	Selected
59	Issac MK (Ed.)	1987a	ICMR-DST, New Delhi Report from West Bengal	ICMR-DST	Selected
60	Jaydeokar S	1997	Indian J Psychiatry	INDMED	
61	Katz MM	1988	Cult Med Psychiatry	MEDLINE	
62	Kenkre AM	2000	Indian J Dent Res.	MEDLINE	
63	Kua EH	1985	Acta Psychiatr Scand	MEDLINE	
64	Kulhara P	1986	Acta Psychiatr Scand	MEDLINE	
65	Kulhara P	1978	Br J Psychiatry	MEDLINE	
66	Kulhara P	2000	Psychopathology	MEDLINE	
67	Kumar PNS	1997	Indian J Psychiatry	INDMED	
68	Lobana A	2001	Acta Psychiatr Scand	MEDLINE	
69	Madhav SM	2001	Indian J Community Med	INDMED	

TABLE A2   *(Continued)*

Sl. No.	Chief investigator	Year of report	Name of journal/Source	Source of location	Remarks
70	Malhotra S	1992	Acta Psychiatr Scand	MEDLINE	
71	McCreadie RG	2002	Arch Gen Psychiatry	MEDLINE	
72	McCreadie RG	2003	Br J Psychiatry	MEDLINE	
73	McCreadie RG	1997	Br J Psychiatry	MEDLINE	
74	McCreadie RG	1996	Br J Psychiatry	MEDLINE	
75	McCreadie RG	2002	Br J Psychiatry	MEDLINE	
76	Mehta P	1985	Indian J Psychiatry	H S	Selected
77	Mojtabai R	2001	Br J Psychiatry	MEDLINE	
78	Murthy GV	1998	. Acta Psychiatr Scand	MEDLINE	
79	Nandi D N	1979	Acta Psychiatr Scand	MEDLINE	Selected
80	Nandi D N	1980	Br J Psychiatry	H S	Selected
81	Nandi DN	2000a	Br J Psychiatry	H S	Selected
82	Nandi DN	2000b	Br J Psychiatry	H S	Selected
83	Nandi DN	1975	Indian J Psychiatry	H S	Selected
84	Nandi DN	1976	Indian J Psychiatry	H S	Selected
85	Nandi DN	1977	Indian J Psychiatry	H S	Selected
86	Nandi DN	1978a	Indian J Psychiatry	H S	Selected
87	Nandi DN	1978b	Indian J Psychiatry	H S	Selected
88	Nandi DN	1980	Indian J Psychiatry	H S	Selected
89	Nandi DN	1992	Indian J Psychiatry	H S	Selected
90	Ndetei DM	1984	Acta Psychiatr Scand	MEDLINE	
91	Not mentioned	2000	Harv Ment Health Lett	MEDLINE	
92	Odutoye K	1999	Int J Geriatr Psychiatry	MEDLINE	
93	Padmavathi R	1987	Indian J Psychiatry	H S	Selected
94	Padmavathi R	1998	Psychol Med	MEDLINE	
95	Patel V	2003	CNS Drugs	MEDLINE	
96	Peet M	1998	Psychiatr Genet	MEDLINE	
97	Peet M	2004	World J Biol Psychiatry	MEDLINE	
98	Pote HL	2002	Ethn Health	MEDLINE	
99	Premarajan KC	1993	Indian J Psychiatry	H S	Selected
100	Raguram R	2004	J Nerv Ment Dis	MEDLINE	
101	Raman AC	1972	J Nerv Ment Dis	MEDLINE	
102	Rao S	1966	Int J Soc Psychiatry	MEDLINE	

TABLE A2   *(Continued)*

Sl. No.	Chief investigator	Year of report	Name of journal/Source	Source of location	Remarks
103	Rao S.	1966	Int J Soc Psychiatry	MEDLINE	
104	Reddy MV	1998	Indian J Psychiatry	INDMED	
105	Reddy YC	1997	Can J Psychiatry	MEDLINE	
106	Sachdeva	1986	Indian J Psychiatry	H S	Selected
107	Salleh MR	1990	Singapore Med J	MEDLINE	
108	Sanua VD	1984	Int J Soc Psychiatry	MEDLINE	
109	Sen B	1984	Indian J Psychiatry	H S	Selected
110	Sethi BB	1977	Am J Psychother	H S	
111	Sethi BB	1972	Indian J Psychiatry	H S	Selected
112	Sethi BB	1978	Indian J Psychiatry	H S	
113	Sethi BB	1967	Indian J Psychiatry	H S	Selected
114	Sethi BB	1973	Am J Psychother	MEDLINE	
115	Sethi BB	1974	Br J Psychiatry	MEDLINE	Selected
116	Shah AV	1980	Indian J Psychiatry	H S	Selected
117	Shaji S	1995	Indian J Psychiatry	H S	Selected
118	Sharma S	2001	Indian J Psychiatry	H S	Selected
119	Sivaramakrishnan K	1994	Acta Psychiatr Scand	MEDLINE	
120	Rama Rao BSS	1976	J Med Genet	MEDLINE	
121	Srinivasa Murthy R	2005	Psychol Med	MEDLINE	
122	Srinivasan TN	2002	Schizophr Bull	MEDLINE	
123	Srinivasan TN	2002	Schizophr Res	MEDLINE	
124	Srinivasan TN	2001	Soc Psychiatry Psychiatr Epidemiol	MEDLINE	
125	Surya NC	1964	Transactions of All India Inst Mental Health	H S	Selected
126	Susser E	1995	Br J Psychiatry	MEDLINE	
127	Susser E	1998	Br J Psychiatry	MEDLINE	
128	Tan CH	1990	Ther Drug Monit	MEDLINE	
129	Thacore VR	1975	Br J Psychiatry	H S	Selected
130	Thara R	2004	Br J Psychiatry	MEDLINE	
131	Thara R	2003	Int J Soc Psychiatry	MEDLINE	
132	Thara R	1997	Soc Psychiatry Psychiatr Epidemiol	MEDLINE	

TABLE A2 *(Continued)*

Sl. No.	Chief investigator	Year of report	Name of journal/Source	Source of location	Remarks
133	Thomas CS	1993	Br J Psychiatry	MEDLINE	
134	Tirupati NS	2004	Aust N Z J Psychiatry	MEDLINE	
135	Tiwari AK	2005	Schizophr Res	MEDLINE	
136	Torrey EF	1987	Br J Psychiatry	MEDLINE	
137	Varma VK	1997	Acta Psychiatr Scand	MEDLINE	
138	Varma VK	1997	Br J Psychiatry	MEDLINE	
139	Varma VK	1996	Psychiatr Q	MEDLINE	
140	Verghese A	1973	Ind J Medical Research	H S	Selected
141	Verghese A	1974	Ind J Medical Research	H S	
142	Wai BH	1999	Gen Hosp Psychiatry	MEDLINE	
143	Yamashita I	1990	Jpn J Psychiatry Neurol	MEDLINE	

HS: Hand Searched.

A.5   PURPOSES AND SAMPLING METHODS OF SELECTED STUDIES

The main purpose along with the population studied and the sampling methods adapted in 37 psychiatry epidemiological studies, where schizophrenia was included as a part or exclusively for schizophrenia disorders, were indicated in Table A3.

TABLE A3   Purpose and Sampling Methods of Schizophrenia Studies Selected for Meta- analysis.

Study No.	Chief investi-gator (year)	Purpose and population studied	Census/Sampling method
1	Surya (1964)	Prevalence rates in a suburb of Pondicherry	Census
2	Sethi (1967)	Prevalence rates and socioeconomic status correlates in an urban area of Lucknow	Representative sample of families
3	Gopinath (1968)	Prevalence rates of a village in Bangalore rural district	Census

TABLE A3   *(Continued)*

Study No.	Chief investigator (year)	Purpose and population studied	Census/Sampling method
4	Dube (1970)	Prevalence rates and biosocial correlates in four rural, four semi-rural villages, and an urban area in and around Agra	Census
5	Elnagar (1971)	Prevalence rates in three caste groups (Brahmins, Kshatriyas, Tribes) in a village with five paras	Random sample of three paras
6	Sethi (1972)	Prevalence rates and socioeconomic status correlates in four villages of Lucknow district	Representative sample of families
7	Verghese (1973)	Prevalence rates and socioeconomic status correlates in Vellore town	Three stage random sampling (wards, streets, houses) of families
8	Sethi (1974)	Prevalence rates and socio-economic status correlates of Lucknow city	Three stage random sampling (constituencies, streets, households) of families
9	Nandi (1975)	Prevalence rates and sociocultural correlates in a village mainly inhabited by Muslim community in West Bengal	Census
10	Thacore (1975)	Prevalence rates and sociocultural correlates in suburb area of Lucknow	Representative sample of families
11	Carstairs (1976)	Prevalence rates of three caste groups (Brahmins, Bunts, Mogers) in a coastal village of Karnataka	Fifty percent random sample of families in the three communities
12	Nandi (1976)	Prevalence rates and 1-year incidence rates in a village mainly inhabited by Muslim community in West Bengal	Census
13	Nandi (1977)	Prevalence rates of two tribal communities and Brahmin community in four villages in tribal belt of West Bengal	Census
14	Agarwal (1978)	Prevalence rates in an urban area of Ahmedabad city	Representative sample of families
15	Nandi (1978a)	Prevalence rates in two villages in West Bengal	Census

TABLE A3   *(Continued)*

Study No.	Chief investi-gator (year)	Purpose and population studied	Census/Sampling method
16	Nandi (1978b)	Prevalence rates and 1-year incidence rates in two villages of West Bengal	Census
17	Nandi (1979)	Prevalence rates of caste Hindu, SC, ST, and Muslim communities in three villages of West Bengal	Census
18	Nandi (1980a)	Prevalence rates in Brahmins, SC, and ST staying in a cluster of 28 villages of West Bengal	Census
19	Nandi (1980b)	Prevalence rates in Brahmins and Tribal families staying in certain villages and all the officers staying in a particular area of city	Representative sample of families
20	Shah (1980)	Prevalence rates and socio-economic status correlates in a geographical area of Ahmedabad city	Representative sample of families
21	Isaac (1980)	Prevalence rates and cost effectiveness of three methods of case findings in a village in Bangalore rural district	Census
22	Bhide (1982)	Prevalence rates in an estate near Bangalore	Census
23	Sen (1984)	Prevalence rates and socioeconomic status (caste groups) correlates in a slum area in Kolkata	Census
24	Mehta (1985)	Prevalence rates and socioeconomic status correlates in a block of 68 villages in Tamil Nadu	Fifty percent of households in 11 randomly selected villages
25	Sachdeva (1986)	Prevalence rates and socioeconomic status correlates in a village near Faridkot in Punjab	Census
26	ICMR (1987a)	Prevalence rates of severe mental morbidity in a PHC in Bangalore rural district	Representative sample of 124 villages of the PHC
27	ICMR (1987b)	Prevalence rates of severe mental morbidity in a PHC in Baroda district	Two sections of the PHC covering 12 villages

TABLE A3   *(Continued)*

Study No.	Chief investigator (year)	Purpose and population studied	Census/Sampling method
28	ICMR (1987c)	Prevalence rates of severe mental morbidity in a PHC with 8 anchals in Kolkata rural district	Three anchals covering 33 villages of mainly Muslim community
29	ICMR (1987d)	Prevalence rates of severe mental morbidity in a PHC in Patiala district	A random sample of 51 villages of mainly Sikh community
30	Padmavathi (1987)	Prevalence rates and demographic variables correlates of schizophrenia disorders in two areas within the catchment area of Government General Hospital, Chennai	Census
31	ICMR (1990)	Prevalence rates and socio-economic status correlates in a catchment area of community rural mental health center in rural Bangalore district	Representative sample of families
32	Nandi (1992)	Prevalence rates in an urbanized tribals in an area of a town and rural tribes in a cluster of villages	Census
33	Premarajan (1983)	Prevalence rates and sociodemographic correlates in an urban community of Pondicherry	Representative families
34	Shaji (1995)	Prevalence rates and socioeconomic correlates of priority psychiatric disorders in a panchayat with ten wards in rural Kerala	Two wards covering 1094 villages
35	Nandi (2000a)	Prevalence rates and socioeconomic status correlates in two villages in different rural districts of West Bengal	Census
36	Nandi (2000b)	Prevalence rates and incidence rates, and socioeconomic status correlates in two villages in different districts of West Bengal	Census
37	Sharma (2001)	Prevalence rates and socio-economic status correlates in a random sample of 24 villages and 12 urban blocks in Goa state	Systematic sample of 384 rural families and 192 urban blocks

A.6   GENERAL INFORMATION OF SCHIZOPHRENIA STUDIES

The general information of selected 37 schizophrenia studies are presented in Table A4. The studies covered a period of 38 years from 1964 to 2001. Nandi has contributed highest number (11) of studies followed by ICMR (5) and Sethi (3). West Bengal has contributed highest number (14) of studies followed by Karnataka (6), Uttar Pradesh (5), Tamilnadu (3), Gujarat (3), Pondicherry (2), Punjab (2), Kerala (1), and Goa (1). Out of 37 studies, 23 are coming from rural areas, 7 are from urban studies, 3 from semi-urban, and 4 are based on mixed domicile of both rural and urban sectors. Based on the available reports of 30 studies, the number of families ranged from 82 to 6038. The average family size was 5.3, and it ranged from 4.0 to 7.5.

TABLE A4   General Information of Schizophrenia Studies Selected for Meta-analysis.

Study No.	Chief investigator	Year of report	State/UT	Locality	No. of families	Average family size
1	Surya	1964	Pondicherry	Semi-urban	532	5.1
2	Sethi	1967	UP	Urban	300	5.8
3	Gopinath	1968	Karnataka	Rural	82	5.2
4	Dube	1970	UP	Mixed	6038	4.9
5	Elnagar	1971	WB	Rural	184	7.5
6	Sethi	1972	UP	Rural	500	5.4
7	Verghese	1973	TN	Semi-urban	539	5.4
8	Sethi	1974	UP	Urban	850	5.3
9	Nandi	1975	WB	Rural	177	6.0
10	Thacore	1975	UP	Semi-urban	500	5.4
11	Carstairs	1976	Karnataka	Rural	344	6.2
12	Nandi	1976	WB	Rural	177	6.1
13	Nandi	1977	WB	Rural	590	4.9
14	Agarwal	1978	Gujarat	Urban	200	5.1
15	Nandi	1978a	WB	Rural	477	4.7
16	Nandi	1978b	WB	Rural	450	5.0
17	Nandi	1979	WB	Rural	609	6.1

TABLE A4   *(Continued)*

Study No.	Chief investigator	Year of report	State/UT	Locality	No. of families	Average family size
18	Nandi	1980a	WB	Rural	815	5.0
19	Nandi	1980b	WB	Mixed	404	4.6
20	Shah	1980	Gujarat	Urban	461	5.9
21	Isaac	1980	Karnataka	Rural	733	5.7
22	Bhide	1982	Karnataka	Rural	...	...
23	Sen	1984	WB	Urban	337	6.4
24	Mehta	1985	TN	Rural	1195	5.0
25	Sachdeva	1986	Punjab	Rural	376	5.3
26	ICMR	1987a	Karnataka	Rural	...	...
27	ICMR	1987b	Gujarat	Rural	...	...
28	ICMR	1987c	WB	Rural	...	...
29	ICMR	1987d	Punjab	Rural	...	...
30	Padmavathi	1987	TN	Urban	...	...
31	ICMR	1990	Karnataka	Rural	...	...
32	Nandi	1992	WB	Mixed	353	4.0
33	Premarajan	1993	Pondicherry	Urban	225	4.7
34	Shaji	1995	Kerala	Rural	1094	4.8
35	Nandi	2000a	WB	Rural	387	5.6
36	Nandi	2000b	WB	Rural	506	6.9
37	Sharma	2001	Goa	Mixed	576	7.0

... Not available; WB: West Bengal; UP: Uttar Pradesh; TN: Tamil Nadu.

A.7   PREVALENCE RATES OF SCHIZOPHRENIA STUDIES

The number of persons surveyed, number of schizophrenia cases, and the prevalence rates per thousand along with its 95% confidence interval (CI) in individual studies are as presented in Table A5. It can be noted that the number of persons surveyed in individual studies have one extremely low size of 423. So far as the prevalence rates are considered, there are two

extremely low prevalence rates of 0.70 and 0.95, and two extremely high
prevalence rates of 14.17 and 7.09.

TABLE A5   Prevalence Rates of Schizophrenia Studies Selected for Meta-analysis.

Study No.	Chief investigator	Year of report	No. of persons	No. of cases	Prevalence rate	95% CI
1	Surya	1964	2731	4	1.46	0.0–2.9
2	Sethi	1967	1733	4	2.31	0.1–4.6
3	Gopinath	1968	423	3	7.09	0.0–15.1
4	Dube	1970	29,468	64	2.17	1.6–2.7
5	Elnagar	1971	1383	6	4.34	0.9–7.8
6	Sethi	1972	2691	3	1.11	0.0–2.4
7	Verghese	1973	2904	5	1.72	0.2–3.2
8	Sethi	1974	4481	11	2.46	1.0–3.9
9	Nandi	1975	1060	3	2.83	0.0–6.0
10	Thacore	1975	2696	5	1.85	0.2–3.5
11	Carstairs	1976	2126	9	4.23	1.5–7.0
12	Nandi	1976	1078	4	3.71	0.1–7.3
13	Nandi	1977	2918	7	2.40	0.6–4.2
14	Agarwal	1978	1019	6	5.89	1.2–10.6
15	Nandi	1978a	2230	9	4.04	1.4–6.7
16	Nandi	1978b	2250	11	4.89	2.0–7.8
17	Nandi	1979	3718	21	5.65	3.2–8.1
18	Nandi	1980a	4053	9	2.22	0.8–3.7
19	Nandi	1980b	1862	10	5.37	2.1–8.7
20	Shah	1980	2712	4	1.48	0.0–2.9
21	Isaac	1980	4203	4	0.95	0.0–1.9
22	Bhide	1982	3135	5	1.60	0.2–3.0
23	Sen	1984	2168	12	5.54	2.4–8.7
24	Mehta	1985	5941	11	1.85	0.8–3.0
25	Sachdeva	1986	1989	4	2.01	0.0–4.0
26	ICMR	1987a	35,548	65	1.83	1.4–2.3

TABLE A5   *(Continued)*

Study No.	Chief investigator	Year of report	No. of persons	No. of cases	Prevalence rate	95% CI
27	ICMR	1987b	39,655	70	1.77	1.4–2.2
28	ICMR	1987c	34,582	71	2.05	1.6–2.5
29	ICMR	1987d	36,595	113	3.09	2.5–3.7
30	Padmavathi	1987	101,229	252	2.49	2.2–2.8
31	ICMR	1990	32,645	60	1.84	1.4–2.3
32	Nandi	1992	1424	1	0.70	0.0–2.1
33	Premarajan	1993	1066	2	1.88	0.0–4.5
34	Shaji	1995	5284	19	3.60	2.0–5.2
35	Nandi	2000a	2183	8	3.67	1.1–6.2
36	Nandi	2000b	3488	10	2.87	1.1–4.6
37	Sharma	2001	4022	57	14.17	10.5–17.8

A.8   QUALITY ASSESSMENT OF SELECTED STUDIES

A checklist has been prepared to measure the quality of the studies, which include all aspects of study design, implementation, and analysis. The checklist items along with their scoring are given in Table A6. The individual items are scored separately and added over all dimensions to give a final score as shown in Table A7.

TABLE A6   Checklist for Quality Assessment of Schizophrenia Studies.

Sl. No.	Particulars	Score	Description
1	Definition of schizophrenia	0	Intuition
		2	Own
		5	WHO/IPSS/RPES
2	Study population	5	General
		0	Specific
3	Survey personnel	3	Includes psychiatrist/neurologist + statistician
		2	Includes psychiatrist/neurologist

TABLE A6 *(Continued)*

Sl. No.	Particulars	Score	Description
		1	Only survey personnel
4	Demography specification	0	Not specified
		5	Specified (0.5 each for age, sex, religion, domicile, education, occupation, income, family type, family size, marital status)
5	Sampling method	5	Census/Multistage/Stratified
		4	Simple random
		3	Systematic
		2	Cluster
		0	Not specific
6	Sample size	5	Completely adequate
		2	Fairly adequate
7	Socioeconomic status assessment	1	Pareek and Trivedi/Kuppuswamy/Prasad
		0	Own
8	Classification system	5	ICD
		3	DSM
		0	Intuition
9	Prevalence specification	1	Age specific
		1	Gender specific
		1	Domicile specific
		1	Occupation specific
		1	Family size specific
		1	Family type specific
		1	Marital status specific
		1	Income specific
		1	Religion/Caste specific
		1	Education specific
		1	Age at onset specific
10	Publication standard	3	International
		2	National/Book
		1	Regional/Dissertation
		0	Reports

Maximum Score: 48.

TABLE A7   Quality Assessment Scores of Schizophrenia Studies.

Study No.	Chief investigator (year)	Total score	Quality characteristics									
			1	2	3	4	5	6	7	8	9	10
1	Surya (1964)	17.0	0	0	2	4.0	5	5	0	0	1	0
2	Sethi (1967)	15.0	0	0	2	5.0	0	5	0	0	1	2
3	Gopinath (1968)	14.0	0	0	2	3.0	5	2	0	0	1	1
4	Dube (1970)	33.0	2	5	3	5.0	5	5	0	0	5	3
5	Elnagar (1971)	16.0	2	0	1	3.0	2	2	0	0	3	3
6	Sethi (1972)	24.0	0	0	2	5.0	0	5	0	0	10	2
7	Verghese (1973)	30.5	5	0	2	4.5	5	5	1	5	1	2
8	Sethi (1974)	26.0	2	0	3	4.0	5	5	0	3	1	3
9	Nandi (1975)	28.5	5	0	3	2.5	5	2	0	5	4	2
10	Thacore (1975)	21.5	2	0	2	4.5	0	5	1	3	1	3
11	Carstairs (1976)	27.5	5	0	2	3.5	4	5	0	5	1	2
12	Nandi (1976)	25.0	5	0	2	2.0	5	2	0	5	2	2
13	Nandi (1977)	32.0	5	0	3	3.0	5	5	1	5	3	2
14	Agarwal (1978)	7.0	0	0	2	1.0	0	2	0	0	1	1
15	Nandi (1978a)	25.0	5	0	2	1.0	5	5	0	5	1	1
16	Nandi (1978b)	25.0	5	0	2	1.0	5	5	0	5	1	1
17	Nandi (1979)	26.0	5	0	2	2.0	5	5	1	0	3	3
18	Nandi (1980a)	31.5	5	0	3	2.5	5	5	1	5	3	2
19	Nandi (1980b)	30.0	5	5	2	3.0	0	5	0	5	3	2
20	Shah (1980)	15.5	0	0	2	4.5	0	5	1	0	1	2
21	Isaac (1980)	27.5	5	0	2	1.5	5	5	0	5	1	3
22	Bhide (1982)	14.5	0	0	2	0.5	5	5	0	0	1	1
23	Sen (1984)	27.5	2	0	3	2.5	5	5	1	5	2	2
24	Mehta (1985)	28.5	5	0	2	2.5	4	5	1	5	2	2
25	Sachdeva (1986)	31.0	5	0	2	4.0	5	5	1	5	2	2
26	ICMR (1987a)	25.0	5	0	3	4.0	2	5	0	5	1	0
27	ICMR (1987b)	25.0	5	0	3	4.0	2	5	0	5	1	0
28	ICMR (1987c)	25.0	5	0	3	4.0	2	5	0	5	1	0

TABLE A7   *(Continued)*

Study No.	Chief investigator (year)	Total score	Quality characteristics									
			1	2	3	4	5	6	7	8	9	10
29	ICMR (1987d)	25.0	5	0	3	4.0	2	5	0	5	1	0
30	Padmavathi (1987)	36.0	5	0	2	4.0	5	5	0	5	8	2
31	ICMR (1990)	24.5	5	0	3	0.5	2	5	0	5	1	3
32	Nandi (1992)	25.5	5	0	2	1.5	5	2	1	5	2	2
33	Premarajan (1983)	22.5	5	0	2	4.5	0	2	0	5	2	2
34	Shaji (1995)	26.5	5	0	2	2.5	2	5	1	5	2	2
35	Nandi (2000a)	30.5	5	0	3	2.5	5	5	1	5	1	3
36	Nandi (2000b)	30.5	5	0	3	2.5	5	5	1	5	1	3
37	Sharma (2001)	31.5	5	5	2	1.5	3	5	0	5	3	2

Note:
1. Definition of cases
2. Study population
3. Survey personnel
4. Demography specification
5. Sampling method
6. Sample size
7. Socioeconomic status assessment
8. Classification system
9. Prevalence specification
10. Publication standard

A.9   RESULTS OF ANALYSIS

A9.1   FOREST PLOT OF SCHIZOPHRENIA STUDIES

By executing the "graph(r)" program under the statistical package STATA Version 8.0 on the basic data for prevalence rates of schizophrenia studies listed chronologically, the requisite Forest plot was obtained as shown in Figure A1. The pooled estimate (2.39) based on DL method is represented by vertical broken line in the plot. The highest prevalence rate of 14.17 (Sharma, 2001) was found to be an extreme value. It can be noted that the areas of the rectangles representing the prevalence rates of seven large studies are bigger and as such clearly utilize their narrow CIs represented by short line joining these rectangles. The lines representing the CIs of four studies have not touched the vertical broken line. They are in the left side of the vertical line indicating that these studies have reported significantly low prevalence rates. Similarly, the lines representing the 95% CI of three studies have not touched the pooled estimate of the prevalence rates. They are in the right side of the vertical line indicating significantly high prevalence rates.

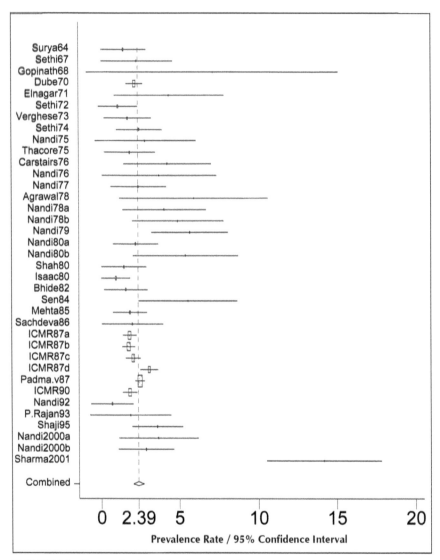

FIGURE A1   Forest plot for schizophrenia studies.

A9.2 FUNNEL PLOT

By executing "metabias" program on the sample sizes and the number of cases in schizophrenia studies, the requisite Begg's Funnel plot was obtained as shown in Figure A2. A perusal of this Funnel plot indicates four studies as heterogeneous as they lie outside the upper CI of the plot. The plot also indicates that the study by Isaac (1980) with a prevalence rate of 0.95 (SE: 0.48) can be considered as heterogeneous.

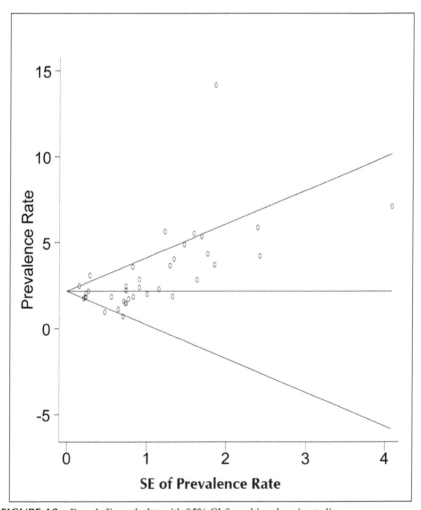

FIGURE A2   Begg's Funnel plot with 95% CI for schizophrenia studies.

A9.3   POOLED ESTIMATES BASED ON FIXED EFFECTS MODEL

There were a total of 962 cases out of a total of 388,693 persons surveyed in these studies, yielding a crude prevalence rate of 2.475 per thousand, with its standard error as 0.032.

In order to calculate the pooled estimate and heterogeneity statistic based on the Inverse-Variance method, the meta-analysis with its basic components of prevalence rate (θ_i), its variance (v_i), its weight (w_i) measured by the precision, the product of the weight and the prevalence rate ($w_i\theta_i$), and finally the product of the square of the deviation of prevalence rates from the pooled estimate and the weight of individual studies are calculated as shown in Table A8. The pooled estimate based on this method is computed as 2.180, and its standard error is computed as 0.075. The heterogeneity statistic (Q_w) for individual studies is shown in the last column and pooled heterogeneity statistic is presented in last row as 115.671 at Table A8. It is more than the chi-square value with 36 degrees of freedom (60.53) and hence significant at 1% level of significance, indicating the significant amount of inter-study variation. This suggests that fixed-effects model meta-analysis was not suitable for prevalence rates of schizophrenia studies.

TABLE A8   Meta-analysis of Schizophrenia Studies—IV Method.

Study No.	Chief investigator (year)	θ_i	v_i	w_i	$w_i\theta_i$	$w_i(\theta_i - \theta_{IV})^2$
1	Surya (1964)	1.465	0.536	1.867	2.735	0.956
2	Sethi (1967)	2.308	1.329	0.753	1.737	0.012
3	Gopinath (1968)	7.092	16.648	0.060	0.426	1.449
4	Dube (1970)	2.172	0.074	13.598	29.532	0.001
5	Elnagar (1971)	4.338	3.123	0.320	1.389	1.491
6	Sethi (1972)	1.115	0.414	2.417	2.694	2.743
7	Verghese (1973)	1.722	0.592	1.690	2.909	0.355
8	Sethi (1974)	2.455	0.546	1.830	4.492	0.138
9	Nandi (1975)	2.830	2.662	0.376	1.063	0.159
10	Thacore (1975)	1.855	0.687	1.456	2.701	0.154
11	Carstairs (1976)	4.233	1.983	0.504	2.135	2.126

TABLE A8   *(Continued)*

Study No.	Chief investigator (year)	θ_i	v_i	w_i	$w_i\theta_i$	$w_i(\theta_i - \theta_{IV})^2$
12	Nandi (1976)	3.711	3.429	0.292	1.082	0.683
13	Nandi (1977)	2.399	0.820	1.219	2.925	0.058
14	Agarwal (1978)	5.888	5.744	0.174	1.025	2.393
15	Nandi (1978a)	4.036	1.803	0.555	2.239	1.910
16	Nandi (1978b)	4.889	2.162	0.462	2.261	3.393
17	Nandi (1979)	5.648	1.511	0.662	3.739	7.962
18	Nandi (1980a)	2.221	0.547	1.829	4.062	0.003
19	Nandi (1980b)	5.371	2.869	0.349	1.872	3.548
20	Shah (1980)	1.475	0.543	1.841	2.716	0.916
21	Isaac (1980)	0.952	0.226	4.421	4.207	6.671
22	Bhide (1982)	1.595	0.508	1.969	3.140	0.674
23	Sen (1984)	5.535	2.539	0.394	2.180	4.433
24	Mehta (1985)	1.852	0.311	3.215	5.952	0.347
25	Sachdeva (1986)	2.011	1.009	0.991	1.993	0.028
26	ICMR (1987a)	1.829	0.051	19.477	35.613	2.408
27	ICMR (1987b)	1.765	0.044	22.504	39.725	3.875
28	ICMR (1987c)	2.053	0.059	16.879	34.653	0.273
29	ICMR (1987d)	3.088	0.084	11.888	36.708	9.794
30	Padmavathi (1987)	2.489	0.025	40.765	101.482	3.898
31	ICMR (1990)	1.838	0.056	17.794	32.705	2.084
32	Nandi (1992)	0.702	0.493	2.029	1.425	4.432
33	Premarajan (1993)	1.876	1.757	0.569	1.068	0.053
34	Shaji (1995)	3.596	0.678	1.475	5.303	2.955
35	Nandi (2000a)	3.665	1.673	0.598	2.191	1.318
36	Nandi (2000b)	2.867	0.820	1.220	3.498	0.576
37	Sharma (2001)	14.172	3.474	0.288	4.080	41.398
Total				178.729	389.658	115.671

$\theta_{IV} = 389.658/178.729 = 2.180$; $SE(\theta_{IV}) = 1/SQRT(178.729) = 0.075$

A9.4 POOLED ESTIMATES BASED ON RANDOM EFFECTS MODEL

In order to proceed with the simplest method of random-effects model of meta-analysis, the DL method, the inter-study variation (τ^2) is computed as 0.5014. The requisite component, namely, the prevalence rate (θ_i), variance (v_i), the adjusted variance ($v_i + \tau^2$), the adjusted weight (w_i'), and the product of the adjusted weight and prevalence rates of individual studies are calculated as shown in Table A9. The pooled estimate based on this method was computed as 2.390, with its standard error as 0.177. Hence 2.390 per thousand population is the pooled estimate of schizophrenia in India.

TABLE A9   Meta-analysis of Schizophrenia Studies—DL Method.

Study No.	Chief investigator (year)	θ_i	v_i	$v_i + \tau^2$	w_i'	$w_i'\theta_i$
1	Surya (1964)	1.465	0.536	1.037	0.964	1.412
2	Sethi (1967)	2.308	1.329	1.830	0.546	1.261
3	Gopinath (1968)	7.092	16.648	17.149	0.058	0.414
4	Dube (1970)	2.172	0.074	0.575	1.739	3.777
5	Elnagar (1971)	4.338	3.123	3.625	0.276	1.197
6	Sethi (1972)	1.115	0.414	0.915	1.093	1.218
7	Verghese (1973)	1.722	0.592	1.093	0.915	1.575
8	Sethi (1974)	2.455	0.546	1.048	0.954	2.343
9	Nandi (1975)	2.830	2.662	3.164	0.316	0.895
10	Thacore (1975)	1.855	0.687	1.188	0.842	1.561
11	Carstairs (1976)	4.233	1.983	2.484	0.403	1.704
12	Nandi (1976)	3.711	3.429	3.931	0.254	0.944
13	Nandi (1977)	2.399	0.820	1.322	0.757	1.815
14	Agarwal (1978)	5.888	5.744	6.246	0.160	0.943
15	Nandi (1978a)	4.036	1.803	2.304	0.434	1.752
16	Nandi (1978b)	4.889	2.162	2.664	0.375	1.835
17	Nandi (1979)	5.648	1.511	2.012	0.497	2.807
18	Nandi (1980a)	2.221	0.547	1.048	0.954	2.119

TABLE A9 *(Continued)*

Study No.	Chief investigator (year)	θ_i	v_i	$v_i + \tau^2$	w_i'	$w_i'\theta_i$
19	Nandi (1980b)	5.371	2.869	3.370	0.297	1.594
20	Shah (1980)	1.475	0.543	1.044	0.957	1.412
21	Isaac (1980)	0.952	0.226	0.728	1.374	1.308
22	Bhide (1982)	1.595	0.508	1.009	0.991	1.580
23	Sen (1984)	5.535	2.539	3.040	0.329	1.821
24	Mehta (1985)	1.852	0.311	0.813	1.231	2.279
25	Sachdeva (1986)	2.011	1.009	1.510	0.662	1.331
26	ICMR (1987a)	1.829	0.051	0.553	1.809	3.308
27	ICMR (1987b)	1.765	0.044	0.546	1.832	3.234
28	ICMR (1987c)	2.053	0.059	0.561	1.784	3.662
29	ICMR (1987d)	3.088	0.084	0.586	1.708	5.273
30	Padmavathi (1987)	2.489	0.025	0.526	1.901	4.733
31	ICMR (1990)	1.838	0.056	0.558	1.793	3.296
32	Nandi (1992)	0.702	0.493	0.994	1.006	0.706
33	Premarajan (1993)	1.876	1.757	2.258	0.443	0.831
34	Shaji (1995)	3.596	0.678	1.179	0.848	3.049
35	Nandi (2000a)	3.665	1.673	2.174	0.460	1.686
36	Nandi (2000b)	2.867	0.820	1.321	0.757	2.170
37	Sharma (2001)	14.172	3.474	3.975	0.252	3.565
Total					31.970	76.408

$\theta_{DL} = 76.408/31.970 = 2.390$; $SE(\theta_{DL}) = 1/SQRT(31.970) = 0.177$

A9.5 POOLED ESTIMATES BASED ON ITERATIVE SCHEMES

The maximum likelihood estimate (MLE), the restricted maximum likelihood estimate (REML) and empirical Bayesian estimate are the iterative schemes for estimating inter-study variation (τ^2). The MLE method is considered when the variance of the estimate is assumed as known. The REML method is an alternative to the MLE method, which leads to unbiased estimate. An estimator for individual studies can be computed by substituting REML estimates for hyper-parameters and this type of approximation to the posterior distribution is used in empirical Bayesian method (EB Method). In the present study, starting with the τ^2 value as given in the DL method, the values of the new pooled estimate and its standard error for each of the three methods are as given in Table A10. These provide for modified weight w^* leading to new estimate for τ^2. The procedure continued until convergence has taken place. The convergence has taken place in the 40th iteration for the ML method, 30th iteration for the REML method, and 29th iteration for the EB method. The respective pooled estimates of prevalence rates of schizophrenia were 2.536, 2.538, and 2.827 with their respective standard errors as 0.224, 0.226, and 0.350, respectively.

TABLE A10 Estimates of Inter-study Variation and Pooled Estimates at Different Iteration Levels of Three Methods for Meta-analysis.

Particulars		ML Method	REML Method	EB Method
Initial estimates				
	τ^2	0.5173	0.5173	0.5173
	θ	2.3963	2.3963	2.3963
	SE (θ)	0.1792	0.1792	0.1792
First iteration				
	τ^2	0.5998	0.6020	1.7638
	θ	2.4231	2.4220	2.6646
	SE (θ)	0.1876	0.1871	0.2711
Second iteration				
	τ^2	0.6745	0.6932	2.7573
	θ	2.4457	2.4484	2.7760
	SE (θ)	0.1947	0.1956	0.3215
Last iteration				
Iteration number		40	30	29
	τ^2	1.0293	1.0715	3.3996
	θ	2.5356	2.5381	2.8272
	SE (θ)	0.2239	0.2259	0.3495

A9.6 RESULTS OF THE BEST METHOD

The pooled estimate was 2.538 for the REML method, whereas it was 2.827 for the EB method. The standard error of the REML estimate (0.226) is less than that of the EB method (0.350) and hence as per the criteria, set for the best method; the result of the REML method was selected. Hence, 2.538 per thousand population is the pooled estimate of schizophrenia in India.

A9.7 RESULTS OF SENSITIVITY ANALYSIS

The prevalence rates according to the items used for the quality assessment are presented in Table A11.

TABLE A11   Sensitivity Meta-analysis of Schizophrenia Studies.

Sl. No.	Methodology	Items	No. of studies	Prevalence rate	Standard error
1	Definition	Intuition	7	1.594	0.332
		Own	5	2.290	0.292
		IPSS/RPES	25	2.848	0.424
2	Study population	General	34	2.213	0.171
		Specific	3	7.076	3.578
3	Survey personnel	Includes psychiatrist + Statistician	13	2.217	0.168
		Includes psychiatrist	23	2.989	0.544
		Only survey personnel	1	4.338	1.767
4	Demographic	Age	27	2.281	0.155
		Sex	33	2.480	0.293
		Religion	20	2.916	0.534
		Domicile	37	2.538	0.226
		Education	18	2.140	0.150
		Occupation	14	2.135	0.158
		Income	22	2.185	0.162
		Family type	12	2.136	0.165
		Family size	26	2.212	0.243

TABLE A11 *(Continued)*

Sl. No.	Methodology	Items	No. of studies	Prevalence rate	Standard error
		Marital status	16	2.130	0.154
5	Sampling method	Census/Multistage/ Stratified	20	2.349	0.273
		Simple random	2	2.691	1.138
		Systematic	1	14.172	1.864
		Cluster	7	2.232	0.268
		Not specific	7	1.827	0.442
6	Sample size	Completely adequate	30	2.534	0.316
		Fairly adequate	7	2.770	0.808
7	Socioeconomic	Pareek & Trivedi/ Kuppuswamy/Prasad	13	2.402	0.359
		Own	24	2.841	0.463
8	Classification system	ICD	25	2.801	0.416
		DSM	2	2.189	0.552
		Intuition	10	2.253	0.474
9	Prevalence	Age specific	5	2.266	0.221
		Gender specific	16	3.344	0.729
		Domicile specific	37	2.538	0.226
		Occupation specific	3	2.186	0.317
		Family size specific	1	1.115	0.643
		Family type specific	2	1.944	0.673
		Marital status specific	3	1.910	0.673
		Income specific	1	1.115	0.643
		Religion/Caste specific	8	4.477	1.411
		Education specific	3	2.186	0.317
		Age at onset specific	2	2.408	0.139
10	Publication standard	International	9	2.371	0.410
		National/Book	18	2.985	0.630
		Regional/Dissertation	5	3.759	1.007
		Reports	5	2.097	0.270

The item prevalence rates which significantly differ from the pooled prevalence rate of 2.538 (SE: 0.226) are specified in the table. The prevalence rates of methodological items which are significantly high were specific to study population, survey personnel without statistician, publication at national or regional level, studies with gender prevalence specification, and with religion/caste specification. Similarly, the prevalence rates of methodological items which are significantly low were general population studies, survey personnel along with statistician, cluster sampling, unpublished reports, demographic specification with socioeconomic factors such as education, occupation and income, and marital status.

The estimated prevalence rate increases as the case definition tightens and thus the prevalence rate was 2.848 for studies in which IPSS or RPES were used. The prevalence rate based on three specific populations was 7.076, which is significant at 1% level of significance. The study by nonpsychiatrist (Elnagar, 1971) yielded a significantly high prevalence rate of 4.338.

The studies with demographic specific prevalence rates were not mutually exclusive and large number of studies involved in each demography specific rates, and consequently their estimates had high precision. Their estimates ranged from 2.130 (marital status specification) to 2.916 (religion specification). The highest prevalence rate of 14.172 in Sharma (2001) study was based on the systematic sampling, which was found significant as shown in the table. The prevalence rate based on completely adequate sample size study was almost identical with the pooled estimate. The studies with standard method of assessing socioeconomic status had low prevalence rate (2.402) as compared to those of fairly adequate sample size studies. The prevalence estimate based on ICD classification (2.801) was more than the prevalence estimate based on DSM classification (2.189).

The studies with prevalence specific rates were not mutually exclusive, but a large number of studies involved in the sex-specific prevalence rates. Besides all the studies were domicile specific. Higher prevalence rate of 4.477 was obtained among religion/caste specific prevalence rates followed by those of sex-specific prevalence rate of 3.344. The prevalence rate of 2.371 based on international publication studies was close as compared to those other studies.

The contents in Table A11 for sensitivity analysis are depicted diagrammatically using Forest plot as shown in Figure A3. It can be viewed that the, CIs of methodological items with large number of studies were narrowed as compared to those methodological items with few number of studies. The prevalence rate based on a systematic sampling method (Sharma, 2001) was isolated from the rest of estimates.

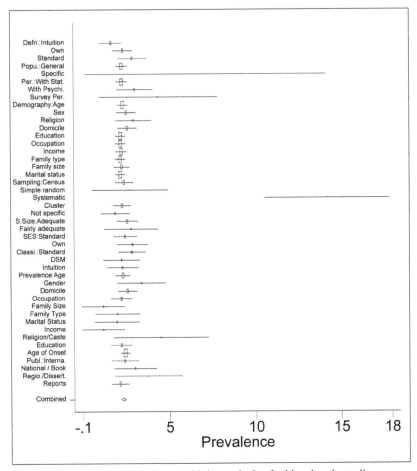

FIGURE A3   Forest plot depicting sensitivity analysis of schizophrenia studies.

A9.8   INFLUENCE ANALYSIS

By executing the program "metainf" with fixed-effects model option, on the basic data of prevalence rates of schizophrenia studies, the influence meta-analysis table as well as the influence meta-analysis plot were obtained as shown in Table A12 and Figure A4, respectively. It can be noted that omitting the Padmavathi (1987) study with a prevalence rate of 2.49 and sample size of 101,229 had the highest influence of lowering the pooled estimate from 2.180 to 2.084, followed by omitting the ICMR (1987d) study with a prevalence rate of 3.09 and sample size of 36,595 from 2.180 to 2.111, omitting the Sharma (2000) study with a prevalence rate of 14.17 and sample size of 4022 from 2.180 to 2.167, and omitting the study Shaji (1995) study with a prevalence rate of 3.60 and sample size of 5284 from 2.180 to 2.164 in that order. Similarly, by omitting of the ICMR (1987b) study with prevalence rate of 1.77 and sample size of 39,655 had highest influence of boosting the pooled estimate from 2.180 to 2.236 followed by omitting the ICMR (1987a) study with prevalence rate of 1.83 and a sample size of 35,548 from 2.180 to 2.219, omitting the ICMR (1990) study with a prevalence rate of 1.84 with sample size of 32,645 from 2.180 to 2.214, and omitting the Isaac (1980) study with a prevalence rate of 0.95 and sample size of 4203 from 2.180 to 2.207 in that order.

TABLE A12   Influence Meta-analysis Table of Schizophrenia Studies.

Study No.	Omitted study	Prevalence rate	Prevalence rate (after)	Lower CI	Upper CI
1	Surya64	1.465	2.184	2.036	2.331
2	Sethi67	2.308	2.176	2.029	2.323
3	Gopinath68	7.092	2.175	2.028	2.321
4	Dube70	2.172	2.177	2.024	2.329
5	Elnagar71	4.338	2.172	2.026	2.319
6	Sethi72	1.115	2.191	2.043	2.339
7	Verghese73	1.722	2.181	2.033	2.328
8	Sethi74	2.455	2.173	2.026	2.321
9	Nandi75	2.830	2.175	2.028	2.322
10	Thacore75	1.855	2.179	2.032	2.326
11	Carstairs76	4.233	2.174	2.028	2.321
12	Nandi76	3.711	2.174	2.027	2.321

TABLE A12 *(Continued)*

Study No.	Omitted study	Prevalence rate	Prevalence rate (after)	Lower CI	Upper CI
13	Nandi77	2.399	2.175	2.028	2.322
14	Agrawal78	5.888	2.173	2.026	2.320
15	Nandi78a	4.036	2.171	2.024	2.317
16	Nandi78b	4.889	2.169	2.022	2.316
17	Nandi79	5.648	2.163	2.016	2.310
18	Nandi80a	2.221	2.176	2.028	2.323
19	Nandi80b	5.371	2.170	2.023	2.317
20	Shah80	1.475	2.184	2.036	2.331
21	Isaac80	0.952	2.207	2.059	2.356
22	Bhide82	1.595	2.183	2.035	2.330
23	Sen84	5.535	2.169	2.022	2.316
24	Mehta85	1.852	2.182	2.034	2.330
25	Sachdeva86	2.011	2.177	2.030	2.324
26	ICMR87a	1.829	2.219	2.063	2.374
27	ICMR87b	1.765	2.236	2.079	2.393
28	ICMR87c	2.053	2.189	2.035	2.343
29	ICMR87d	3.088	2.111	1.959	2.263
30	Padma.v87	2.489	2.084	1.917	2.251
31	ICMR90	1.838	2.214	2.059	2.368
32	Nandi92	0.702	2.193	2.046	2.341
33	P. Rajan93	1.876	2.177	2.030	2.324
34	Shaji95	3.596	2.164	2.017	2.312
35	Nandi2000a	3.665	2.171	2.024	2.318
36	Nandi2000b	2.867	2.172	2.024	2.318
37	Sharma2001	14.172	2.157	2.010	2.304

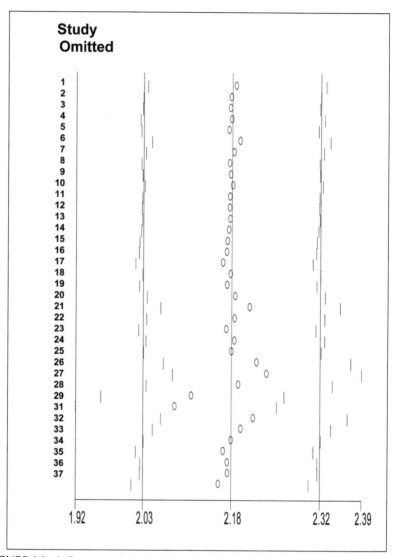

FIGURE A4   Influence analysis plot for schizophrenia studies.

A9.9   RESULTS OF SUBGROUP META-ANALYSIS

The analysis of different categories of variables included in the schizo-phrenia studies are carried out in order to deal with the pattern of preva-lence of schizophrenia in India as shown in Table A13.

TABLE A13   Prevalence Rates of Schizophrenia and Biosocial Correlates.

Sl. No.	Characteristics (number of studies)	Prevalence rate (REML)	Sl. No.	Characteristics (number of studies)	Prevalence rate (REML)
1	Domicile (37)		11	Family type (2)	
	Rural	2.31		Nuclear	1.85
	Urban	2.08		Joint	1.68
	Semi-urban/Mixed	4.62		Extended	3.67
2	Region (37)			Living alone	24.39
	Northern	2.27	12	Family size (1)	
	Eastern	3.18		Up to 5	0.98
	Western	5.63		Above 5	1.20
	Southern	1.98	13	Domicile & sex (17)	
3	Sex (16)			Rural	
	Male	2.29		Males	1.80
	Female	2.60		Females	2.35
4	Age (4)			Urban	
	0–	–		Males	1.06
	5–	0.12		Females	2.22
	10–	3.58		Mixed domicile	
	20–	3.83		Males	11.57
	30–	6.12		Females	4.30
	40–	3.94	14	Caste & sex (3)	
	50–	3.24		Brahmins	
5	Marital status (3)			Males	10.23
	Single	0.33		Females	5.68
	Married	2.79		Scheduled castes	
6	Religion (3)			Males	2.32
	Hindu	5.14		Females	0.06
	Muslim	1.69		Scheduled tribes	
	Christian	4.74		Males	0.12
	All others	0.07		Females	1.90
7	Caste (3)		15	Age of onset & sex (1)	
	Brahmins	7.46		0–4 age	

TABLE A13   *(Continued)*

Sl. No.	Characteristics (number of studies)	Prevalence rate (REML)	Sl. No.	Characteristics (number of studies)	Prevalence rate (REML)
	SC	2.77		Males	–
	ST	0.88		Females	–
	All others	1.55		5–14 age	
8	Literacy level (3)			Males	0.90
	Illiterate/Primary	3.00		Females	–
	Secondary	1.40		15–24 age	
	University	0.68		Males	5.46
9	Occupation (3)			Females	4.31
	Agriculture	3.60		25–34 age	
	Laborers	3.20		Males	4.25
	Professional/ Executive	1.82		Females	3.73
	Business	3.42		35–44 age	
	Retired	26.32		Males	3.04
	Unemployed	23.56		Females	4.39
10	Income (1)			45–54 age	
	Low	1.63		Males	2.53
	Middle	1.04		Females	2.06
	High	–		54 + age	
				Males	0.85
				Females	–

A9.10   RESULTS OF CUMULATIVE META-ANALYSIS

The cumulative meta-analysis was employed in the present study on the prevalence rates of schizophrenia studies in order to determine the trend of these rates during the period. Table A14 was prepared with their columns for identification of the study and year of report, number of persons studied, number of cases, the prevalence rate, the cumulative prevalence rate based on the DL method of meta-analysis, and their 95% CI.

By executing the "metacum" program with DL method meta-analysis option, under the statistical package STATA Version 8.0 on a minicomputer with the Intel CPU 80486 under MS Windows operating system, on the basic information for the prevalence rates of 37 schizophrenia studies listed chronologically, the requisite cumulative plot was obtained as shown in Figure A5. A broad visual examination of the cumulative plot for schizophrenia studies may suggest a considerable increasing trend during the period of four decades. However, a cyclic variation is observed: the prevalence rates increases up to the 19th study and then decreases up to the 28th study and maintaining the same up to the 36th study. The last study by Sharma (2001) with prevalence rate of 14.17 had considerable decreasing impact. The study by Sethi (1972) with a prevalence rate of 1.11 had a considerable decreasing impact, while the studies by Nandi (1979) with a prevalence rate of 5.65 and ICMR (1978d) with a prevalence rate of 3.09 had sudden increasing impact. During the initiating period, the length of the CI narrowed when the major study by Dube (1970) with a sample size of 29,468 and prevalence rate of 2.17 had reached.

TABLE A14   Cumulative Meta-analysis of Schizophrenia Studies.

Study No.	Chief investigator (year)	No. of persons	No. of cases	Prevalence rate	Cumulative prevalence rate	95% CI
1	Surya (1964)	2731	4	1.46	1.47	0.03–2.90
2	Sethi (1967)	1733	4	2.31	1.71	0.50–2.92
3	Gopinath (1968)	423	3	7.09	1.85	0.59–3.11
4	Dube (1970)	29,468	64	2.17	2.12	1.63–2.60
5	Elnagar (1971)	1383	6	4.34	2.16	1.68–2.64
6	Sethi (1972)	2691	3	1.11	1.96	1.31–2.61
7	Verghese (1973)	2904	5	1.72	1.97	1.49–2.46
8	Sethi (1974)	4481	11	2.46	2.04	1.63–2.45
9	Nandi (1975)	1060	3	2.83	2.05	1.64–2.46
10	Thacore (1975)	2696	5	1.85	2.04	1.64–2.44
11	Carstairs (1976)	2126	9	4.23	2.08	1.69–2.48
12	Nandi (1976)	1078	4	3.71	2.10	1.71–2.49
13	Nandi (1977)	2918	7	2.40	2.12	1.73–2.50

TABLE A14 *(Continued)*

Study No.	Chief investigator (year)	No. of persons	No. of cases	Prevalence rate	Cumulative prevalence rate	95% CI
14	Agarwal (1978)	1019	6	5.89	2.14	1.76–2.52
15	Nandi(1978a)	2230	9	4.04	2.19	1.78–2.61
16	Nandi (1978b)	2250	11	4.89	2.31	1.82–2.80
17	Nandi (1979)	3718	21	5.65	2.58	1.99–3.17
18	Nandi (1980a)	4053	9	2.22	2.52	1.98–3.06
19	Nandi (1980b)	1862	10	5.37	2.62	2.06–3.18
20	Shah (1980)	2712	4	1.48	2.52	1.99–3.05
21	Isaac (1980)	4203	4	0.95	2.41	1.88–2.95
22	Bhide (1982)	3135	5	1.60	2.34	1.84–2.85
23	Sen (1984)	2168	12	5.54	2.45	1.93–2.97
24	Mehta (1985)	5941	11	1.85	2.38	1.89–2.86
25	Sachdeva (1986)	1989	4	2.01	2.35	1.88–2.82
26	ICMR (1987a)	35,548	65	1.83	2.24	1.83–2.65
27	ICMR (1987b)	39,655	70	1.77	2.15	1.79–2.51
28	ICMR (1987c)	34,582	71	2.05	2.11	1.79–2.44
29	ICMR (1987d)	36,595	113	3.09	2.24	1.89–2.58
30	Padmavathi (1987)	101,229	252	2.49	2.24	1.93–2.56
31	ICMR (1990)	32,645	60	1.84	2.21	1.91–2.50
32	Nandi (1992)	1424	1	0.70	2.17	1.87–2.46
33	Premarajan (1983)	1066	2	1.88	2.16	1.87–2.45
34	Shaji (1995)	5284	19	3.60	2.20	1.91–2.49
35	Nandi (2000a)	2183	8	3.67	2.21	1.93–2.50
36	Nandi (2000b)	3488	10	2.87	2.23	1.94–2.51
37	Sharma (2001)	4022	57	14.17	2.39	2.04–2.74

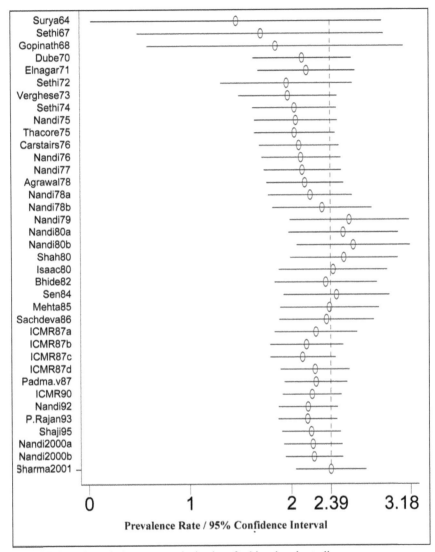

FIGURE A5   Cumulative meta-analysis plot of schizophrenia studies.

A9.11   RESULTS OF META-REGRESSION ANALYSIS

An attempt has been made to employ meta-regression analysis in the present study to identify relationship between the prevalence rates (dependent variable) and the quality assessment scores of studies (independent variable). The REML estimate of between-study variance (τ^2) is computed as 1.305, the constant as 1.657 (95% CI: −1.268–4.582) and the regression coefficient of 0.036 (95% CI: −0.075–0.147). From Figure A6, it is configured that there is a linear fit indicting the positive relationship between the prevalence rates and the quality assessment scores. The lowest score of 7.0 (Agarwal, 1978) is followed by score 14.0 (Gopinath, 1968). The largest score of 36.0 (Padmavathi, 1987) was just before the linear trend line as shown in Figure A6, while the Sharma (2001) study was isolated from the linear trend.

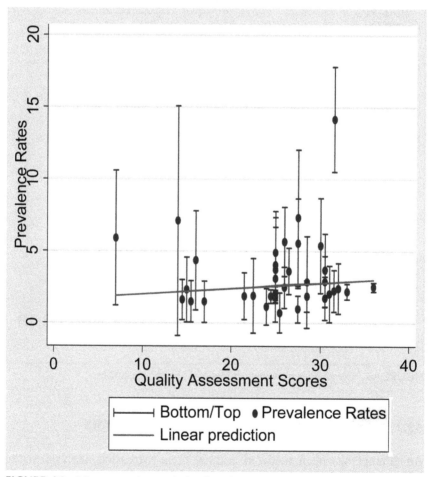

FIGURE A6   Meta-regression analysis plot of prevalence rates on quality assessment scores for schizophrenia studies.

A9.12   RESULTS OF CLUSTER META-ANALYSIS

An attempt has been made to incorporate clustering methods with meta-analytical approach to deal with the pattern of prevalence of schizophrenia in the present study. Two hierarchical agglomerative methods, namely, the complete linkage method and the average linkage between groups were employed to cluster the schizophrenia studies on the quality assessment scores along with domicile and regional information. Each of these two methods was employed with both Euclidian distance and absolute Euclidian distance. The Rand index to determine the level of agreement among the clusters, and c-index to determine the number of clusters present in the data was also employed. The statistical software "Mechaon CVE" was used to carry out the above cited cluster analysis procedures.

All the four Dendrograms (two methods and two proximity measures) obtained had clear cut presence of clusters. The cluster solution of average linkage between groups using Euclidian distance had the highest agreement as judged by the Rand index values. Further, the c-index values were lowest at four number of clusters indicating the presence of four clusters in the data. Empirical investigation also supports the presence of four clusters. Hence, the 4-cluster solution of average linkage between groups with Euclidian distance was considered as the best solution to determine the pattern of prevalence of schizophrenia in the present study. The Dendrogram of this method is shown in Figure A7.

A9.13   CONCLUSION

Meta-analysis is clearly superior to individual studies, traditional narrative studies and systematic reviews in estimating prevalence rates of schizophrenia as employed in the present study. Meta-analytic subgroup analysis with available data provides more realistic approach to determine the pattern of prevalence of schizophrenia as obtained in the present study.

The present study concluded that it is possible to clearly state that meta-analysis of observational studies such as the present study will be as successful as meta-analysis of randomized controlled studies.

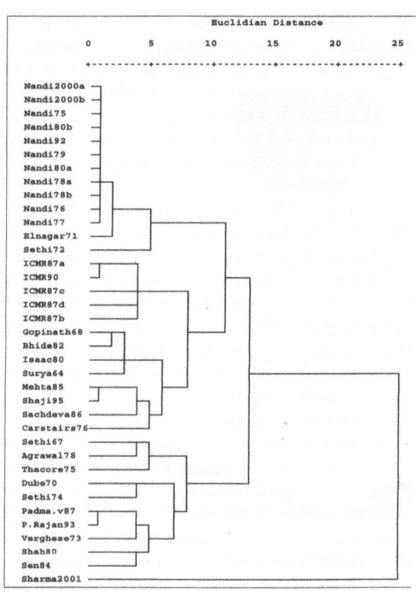

FIGURE A7 Dendrogram using average linkage between groups for schizophrenia studies.

REFERENCES

Agarwal, R. B. Epidemiological Study of Psychiatric Disorders in Urban Setting (200 Families), M.D. Thesis, Gujarat University, 1978; pp 37–38.

Bhide, A. Prevalence of Psychiatric Morbidity in a Closed Community in South India, M.D. Thesis, National Institute of Mental Health and Neuro Sciences, Bangalore, 1982.

Carstairs, G. M.; Kapur, R. L. The Great Universe of Kota—Stress Change and Mental Disorder in an Indian Village. Hogarth Press: London, 1976, 89–114.

Dube, K. C. A Study of Prevalence and Biosocial Variables in Mental Illness in a Rural and an Urban Community in Uttar Pradesh, India. *Acta Psychiatr. Scand.* **1970**, *46*, 327–359.

Elnagar, M. N.; Maitra, P.; Rao, M. N. Mental Health in an Indian Rural Community. *Br. J. Psychiatry* **1971**, *118*, 499–503.

Gopinath, P. S. Epidemiology of Mental Illness in an Indian Village. M.D. Thesis, Bangalore University: Bangalore, 1968; pp 54–71.

Indian Council of Medical Research Report. In *Collaborative Study on Severe Mental Morbidity*; Issac, M. K., Ed.; Indian Council of Medical Research and Department of Science and Technology: New Delhi, 1987a,b,c,d.

Indian Council of Medical Research Report. *Prevalence of Mental Disorders in Adult Rural Population of Bangalore Urban District*; Srinivas Murthy, et al.; 1990, Unpublished data.

Issac, M. K.; Kapur, R. L. A Cost-effectiveness Analysis of Three Different Methods of Psychiatric Case Finding in the General Population. *Br. J. Psychiatry* **1980**, *137*, 540–546.

Mehta, P.; Joseph, A.; Verghese, A. An Epidemiological Study of Psychiatric Disorders in a Rural Area in Tamilnadu. *Indian J. Psychiatry* **1985**, *27*, 153–158.

Nandi, D. N.; Ajmany, S.; Ganguli, H.; Banerjee, G.; Boral, G. C.; Ghosh, A.; Sarkar, S. The Incidence of Mental Disorders in One Year in a Rural Community in West Bengal. *Indian J. Psychiatry* **1976**, *18*, 79–87.

Nandi, D. N.; Ajmany, S.; Ganguly, H.; Banerjee, G.; Boral, G. C.; Ghosh, A. Psychiatric Disorders in a Rural Community in West Bengal: An Epidemiological Study. *Indian J. Psychiatry* **1975**, *17*, 87–99.

Nandi, D. N.; Banerjee, G.; Boral, G. C.; Ganguli, H.; Ajmany, S.; Ghosh, A.; Sarkar, S. Socio-Economic Status and Prevalence of Mental Disorders in Certain Rural Communities in India. *Acta Psychiatr. Scand.* **1979**, *59*, 276–293.

Nandi, D. N.; Banerjee, G.; Chowdhury, A. N.; Banerjee, T.; Boral, G. C.; Sen, B. Urbanisation and Mental Morbidity in Certain Tribal Communities in West Bengal. *Indian J. Psychiatry* **1992**, *34*, 334–339.

Nandi, D. N.; Banerjee, G.; Ganguli, H.; Ajmany, S.; Boral, G. C.; Ghosh, A.; Sarkar, S. The Natural History of Mental Disorders in a Rural Community: A Longitudinal Field Survey. *Indian J. Psychiatry* **1978b**, *21*, 390–396.

Nandi, D. N.; Banerjee, G.; Mukherjee, S. P.; Ghosh, A.; Nandi, P. S.; Nandi, S. Psychiatric Morbidity of a Rural Indian Community—Changes Over a 20-Year Interval. *Br. J. Psychiatry* **2000a,b**, *176*, 351–356.

Nandi, D. N.; Das, N. N.; Chaudhuri, A.; Banerjee, G.; Datta, P.; Ghosh, A.; Boral, G. C. Mental Morbidity and Urban Life: An Epidemiological Study. *Indian J. Psychiatry* **1980b**, *22*, 324–330.

Nandi, D. N.; Mukherjee, S. P.; Banerjee, G.; Boral, G. C.; Ghosh, A.; Sarkar, S.; Das, S.; Banerjee, K.; Ajmany, S. Psychiatric Morbidity in an Uprooted Community in Rural West Bengal. *Indian J. Psychiatry* **1978a,** *20,* 137–142.

Nandi, D. N.; Mukherjee, S. P.; Boral, G. C.; Banerjee, G.; Ghosh, A.; Ajmany, S.; Sarkar, S.; Biswas, D. Prevalence of Psychiatric Morbidity in Two Tribal Communities in Certain Villages of West Bengal: A Cross Cultural Study. *Indian J. Psychiatry* **1977,** *19,* 2–12.

Nandi, D. N.; Mukherjee, S. P.; Boral, G. C.; Banerjee, G.; Ghosh, A.; Sarkar, S.; Ajmany, S. Socio-economic Status and Mental Morbidity in Certain Tribes and Castes in India: A Cross-cultural Study. *Br. J. Psychiatry* **1980a,** *136,* 73–85.

Padmavathi, R.; Rajkumar, S.; Kumar, N.; Manoharan, A.; Kamath, S. Prevalence of Schizophrenia in an Urban Community in Madras. *Indian J. Psychiatry* **1987,** *31,* 233–239.

Premarajan, K. C.; Danabalan, M.; Chandrasekar, R.; Srinivas, D. K. Prevalence of Psychiatric Morbidity in an Urban Community of Pondicherry. *Indian J. Psychiatry* **1993,** *35,* 99–102.

Sachdeva, J. S.; Singh, S.; Sidhu, B. S.; Goyal, R. K. D.; Singh, J. An Epidemiological Study of Psychiatric Disorders in Rural Faridkot (Punjab). *Indian J. Psychiatry* **1986,** *28,* 317–323.

Sen, B.; Nandi, D. N.; Mukherjee, S. P.; Mishra, D. C.; Banerjee, G.; Sarkar, S. Psychiatric Morbidity in an Urban Slum-Dwelling Community. *Indian J. Psychiatry* **1984,** *26,* 185–193.

Sethi, B. B.; Gupta, S. C.; Mahendru, R. K.; Kumari, P. Mental Health and Urban Life: A Study of 850 Families. *Br. J. Psychiatry* **1974,** *124,* 243–246.

Sethi, B. B.; Gupta, S. C.; Rajkumar, Kumari, P. A Psychiatric Survey of 500 Rural Families. *Indian J. Psychiatry* **1972,** *14,* 183–196.

Sethi, B. B.; Gupta, S. C.; Rajkumar, S. Three Hundred Urban Families: A Psychiatric Study. *Indian J. Psychiatry* **1967,** *9,* 280–302.

Shah, A. V.; Goswami, U. A.; Maniar, R. C.; Hajariwala, D. C.; Singh, B. K. Prevalence of Psychiatric Disorders in Ahmedabad: An Epidemiological Study. *Indian J. Psychiatry* **1980,** *22,* 384–389.

Shaji, S.; Verghese, A.; Promodu, K.; George, B.; Shibu, V. P. Prevalence of Priority Psychiatric Disorders in a Rural Area in Kerala. *Indian J. Psychiatry* **1995,** *37,* 91–96.

Sharma, S.; Singh, M. M. Prevalence of Mental Disorders: An Epidemiological Study in Goa. *Indian J. Psychiatry* **2001,** *43,* 118–126.

Surya, N. C.; Datta, S. P.; Gopalakrishna, R.; Sundaram, D.; Kutty, J. Mental Morbidity in Pondicherry (1962–1963). *Trans. All Indian Inst. Ment. Health (NIMHANS),* **1964,** *4,* 50–61.

Thacore, V. R.; Gupta, S. C.; Suraiya, M. Psychiatric Morbidity in a North Indian Community. *Br. J. Psychiatry* **1975,** *126,* 364–369.

Verghese, A.; Beig, A.; Senseman, L. A.; Rao, S. S. S.; Benjimin, V. A Social and Psychiatric Study of a Representative Group of Families in Vellore Town. *Indian J. Med. Res.* **1973,** *61,* 608–620.

LIST OF ABBREVIATIONS OF LOCATED SOURCES/JOURNALS

Sl. No.	Abbreviations for Journal	Title of Journal
1	Acta Paediatr Jpn	ACTA Paediatrica Japonica (Tokyo)
2	Acta Psychiatr Scand	ACTA Psychiatrica Scandinavica (Copenhagen)
3	Acta Trop	ACTA Tropica (Amsterdam)
4	Am J Prev Med	American Journal of Preventive Medicine (New York, NY)
5	Am J Psychother	American Journal of Psychotherapy (Jamaica, NY)
6	An Med Interna	Anales de Medicina Interna (Madrid)
7	Ann Acad Med Singapore	Annals of the Academy of Medicine, Singapore (Singapore)
8	Ann Trop Med Parasitol	Annals of Tropical Medicine and Parasitology (London)
9	Ann Trop Paediatr	Annals of Tropical Paediatrics (Abingdon)
10	Arch Gen Psychiatry	Archives of General Psychiatry (Chicago, IL)
11	Arq Neuropsiquiatr	Arquivos de Neuro-Psiuiatria (Sao Paulo)
12	Aust N Z J Psychiatry	Australian and New Zealand Journal of Psychiatry (New South Wales)
13	BMJ	BMJ (London)
14	Br J Gen Pract	British Journal of General Practice (London)
15	Br J Psychiatry	British Journal of Psychiatry (London)
16	Bull Soc Pathol Exot	Bulletin de la Societe de Pathologie Exotique (Paris)
17	Bull World Health Organ	Bulletin of the World Health Organization (Geneva)
18	Can J Psychiatry	Canadian Journal of Psychiatry. Revue Canadienne de Psychiatrie (Ottawa)
19	CNS Drugs	CNS Drugs
20	Cult Med Psychiatry	Culture, Medicine and Psychiatry (Dordrecht)
21	Eur Arch Psychiatry Clin Neurosci	European Archives of Psychiatry and Clinical Neuroscience (Berlin)
22	Gen Hosp Psychistry	General Hospital Psychiatry (New York, NY)
23	Indian J Community Med	Indian Journal of Community Medicine.
24	Indian J Med Sci	Indian Journal of Medical Sciences (Bombay)

Sl. No.	Abbreviations for Journal	Title of Journal
25	Indian J of Dermatology	Indian Journal of Dermatology
26	Indian J Pediatr	Indian Journal of Pediatrics (New Delhi)
27	Indian J Psychiatry	Indian Journal of Psychiatry
28	Indian Pediatr	Indian Pediatrics (New Delhi)
29	Int J Epidemiol	International Journal of Epidemiology (London)
30	Int J Soc Psychiatry	International Journal of Social Psychiatry (London)
31	J Assoc Physicians India	Journal of the Association of Physicians of India (Bombay)
2	J Biosoc Sci	Journal of Biosocial Science (Cambridge, Eng)
33	J Child Psychol Psychiatry	Journal of Child Psychology and Psychiatry and Allied Disciplines (Oxford)
34	J Clin Pharmacol	Journal of Clinical Pharmacology (Hagerstown, MD)
35	J Fam Pract	Journal of Family Practice (East Norwalk, CT)
36	J Indian Med Assoc	Journal of the Indian Medical Association (Calcutta)
37	J Med Genet	Journal of Medical Genetics (London)
38	J Med Liban	Journal Medical Libanais. Lebanese Medical Journal (Beirut)
39	J Postgrad Med	Journal of Postgraduate Medicine (Bombay)
40	J R Soc Med	Journal of the Royal Society of Medicine (London)
41	J Trop Med Hyg	Journal of Tropical Medicine and Hygiene (Oxford)
42	J Trop Pediatr	Journal of Tropical Pediatrics (London)
43	Jpn J Psychiatry Neurol	Japanese Journal of Psychiatry and Neurology (Tokyo)
44	Med J Aust	Medical Journal of Australia (Sydney)
45	Med J Malaysia	Medical Journal of Malaysia (Kuala Lumpur)
46	Occup Med	Occupational Medicine (Oxford)
47	Psychiatr Q	Psychiatric Quarterly (New York, NY)
47	Psychiatry	Psychiatry (New York, NY)
49	Psychol Med	Psychological Medicine (London)
50	Psychopathology	Psychopathology (Basel)
51	Schizophr Bull	Schizophrenia Bulletin (Rockville, MD)

Sl. No.	Abbreviations for Journal	Title of Journal
52	Schizophr Res	Schizophrenia Research (Amsterdam)
53	Soc Psychiatry Psychiatr Epidemiol	Social Psychiatry and Psychiatric Epidemiology
54	Southeast Asian J Trop Med Public Health	Southeast Asian Journal of Tropical Medicine and Public Health
55	Ther Drug Monit	Therapeutic Drug Monitoring (New York, NY)
56	Tidsskr Nor Laegeforen	Tidsskrift For Den Norske Laegeforening (Oslo)
57	Trop Doct	TROPICAL DOCTOR (LONDON)
58	Trop Geogr Med	TROPICAL AND GEOGRAPHICAL MEDICINE (HAGUE)
59	Trop Med Int Health	Tropical Medicine & International Health (Oxford)

INDEX